Building Department Administration

3rd Edition

WORKBOOK

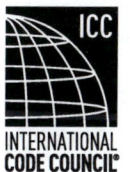

Building Department Administration Workbook

ISBN: 978-1-58001-793-0

Cover Art Director:	Dianna Hallmark
Cover Design:	Duane Acoba
Manager of Development:	Hamid Naderi
Project Editor:	Roger Mensink
Publications Manager:	Mary Lou Luif
Typesetting:	Sue Brockman

COPYRIGHT © 2009

INTERNATIONAL
CODE COUNCIL®

Errata on various ICC publications may be available at www.iccsafe.org/errata.

First Printing: June 2009

PRINTED IN THE U.S.A.

Table of Contents

Chapter 1: The Purpose of Controls. 1

   Chapter 1 Quiz · 3

Chapter 2: Building Codes and Federal Influences 9

   Chapter 2 Quiz · 10

Chapter 3: The Legacy Model Code Groups, Their Codes and the
             Roots of the International Code Council. 17

   Chapter 3 Quiz · 19

Chapter 4: Construction Codes and Standards. 25

   Chapter 4 Quiz · 27

Chapter 5: Building Regulations Around the World 33

   Chapter 5 Quiz · 35

Chapter 6: The Department 41

   Chapter 6 Quiz · 43

Chapter 7: The Development Permit Process 49

   Chapter 7 Quiz · 51

Chapter 8: The Building Official. 57

   Chapter 8 Quiz · 59

Chapter 9: The Effective Manager 65

   Chapter 9 Quiz · 66

Chapter 10: Supervision and Training. 73

   Chapter 10 Quiz · 75

Chapter 11: Department Staffing Requirements. 81

   Chapter 11 Quiz · 83

Chapter 12: The Public Counter. 89

   Chapter 12 Quiz · 91

Chapter 13: Using Information Technology in Building Departments 97

   Chapter 13 Quiz · 99

Chapter 14: Records Management. 105

   Chapter 14 Quiz · 106

Chapter 15: Customer Relations 113

   Chapter 15 Quiz · 115

Chapter 16: Legal Aspects of Code Administration 121

Chapter 16 Quiz · 123

Chapter 17: Disaster Mitigation and Building Security 129

Chapter 17 Quiz · 131

Chapter 18: Housing, Property Maintenance and Code Enforcement Inspection Programs. . . 137

Chapter 18 Quiz · 139

Chapter 19: Building Sustainability: Preserving the Existing Residential Stock 145

Chapter 19 Quiz · 147

Chapter 20: Rehabilitation and General Building Code Approaches 153

Chapter 20 Quiz · 154

Chapter 21: Green Building and Sustainability 157

Chapter 21 Quiz · 158

Answer Keys · 165

Introduction

This *Workbook* to the *Building Department Administration* (BDA), 3rd Edition, is intended to provide practical learning assignments for independent study of the BDA. The independent study format provides a method for the learner to complete the study program in an unregulated time period. Proceeding through the workbook, the student can measure his or her level of knowledge by using the quizzes in each study session.

All study sessions contain specific learning objectives, a lesson note that gives the total perspective of the study session and a list of statements and questions summarizing the key points for study. The quizzes are designed to assess the student's retention of technical knowledge; therefore, before beginning the quizzes, it would be of great benefit if students thoroughly review the corresponding chapter of the BDA and the lesson note and key points of this workbook.

The quizzes are designed to encourage the student to develop the habit of carefully reading the text for a clear understanding of the subject material. The questions are not intended to be tricky or misleading. The following three formats are used to vary the method of evaluation:

1. Multiple Choice: Each statement is followed by a unique group of possible responses from which to choose

2. True/false: Each statement is either true or false

3. Completion: Each statement must be correctly completed by inserting the proper word or text from the BDA

The workbook is structured so that every question is followed by the opportunity to record the learner's response and the corresponding text reference. The correct responses are provided at the back of the workbook in the answer key so that the student can assess his or her knowledge immediately.

It should be noted that the information in this workbook is based on the parent document the *Building Department Administration* (BDA), 3rd Edition, first printing. This is important because some issues in the industry might have evolved, and the current practices might not be exactly the same as reflected in the parent document. As an example, in 2009 the International Code Council® (ICC®) revised its code development process, which is different than the process described in the BDA because the BDA was printed in June of 2007.

Acknowledgements

This workbook was developed by Susan Gentry, Code Consultant with the State of North Carolina office of Fire Marshal; Susan Kelley, Superintendent of Permits and Inspections for the City of Omaha, Nebraska Planning Department; and Maureen Traxler, Code Development Supervisor for the City of Seattle, Washington Department of Planning and Development. Susan Gentry was responsible for Chapters 1 through 10, Susan Kelley was responsible for Chapters 11 through 15 and 17 through 21 and Maureen Traxler was responsible for Chapter 16. All three were also peer reviewers for the *Building Department Administration* (BDA), 3rd Edition. The International Code Council extends its appreciation to the efforts of these individuals for their outstanding work.

Chapter 1
The Purpose of Controls

OBJECTIVES: To become familiar with the historical background of codes and why building regulation is necessary.

LESSON NOTES: This chapter provides a historical perspective of building regulation development. Throughout history, leaders have implemented building regulations, often by using information from tragic events. This chapter broadly outlines how building codes are developed and administered today by the International Code Council® (ICC®). ICC Evaluation Service® (ICC-ES®), which allows for innovative products and components, and International Accreditation Service® (IAS®) are introduced, and their relevance and value to the code enforcement official is discussed.

KEY POINTS:
- What major worldwide historical events led to the earliest historical building regulations?
- What are the two thoughts on why fire losses have decreased over time?
- What is one important point for a building official to make when securing support for building regulation?
- What were some of the specific U.S. tragedies that led to building regulation development in America?
- What types of building regulations were among the earliest in the U.S.?
- How have advances in construction in the past 100 years affected building regulation?
- The early development of what organizations led to better evaluation of building materials and construction methods?
- When was the first recommended building code developed in the U.S.?
- What is the earliest known code?
- What is the primary intent of building regulation?
- What should codes be based on?
- What is the authority for enforcing codes based on?
- What are some of the difficulties encountered when attempting to adequately staff a building department? From what do these difficulties arise?
- What is the result of an inadequately staffed building department?
- What led to the development of the modern building department?

- What types of tragedies tend to lead to building regulation scrutiny today?
- What other influences affect construction regulation?
- How can conflicting jurisdictions and interests negatively affect builders and the public?
- Why is ICC-ES of value to the code official?
- In general, what is the process for an evaluation by ICC-ES?
- What service does the IAS provide?
- What is the process and basis of IAS evaluation for accrediting?
- What sets IAS apart from other accreditation services?
- Why is there a need for ILAC (International Laboratory Accreditation Cooperation)?

Quiz

Chapter 1

Multiple Choice

1. Authority for codes is based on _____.

    a. voting referendums

    b. code enforcement officials' opinion

    c. police power

    d. engineering standards

    Reference _____

2. The earliest recommended U.S. building code was issued by_____.

    a. Building Officials Conference of America (BOCA)

    b. Southern Building Code Congress International (SBCCI)

    c. National Board of Fire Underwriters

    d. International Conference of Building Officials (ICBO)

    Reference _____

3. The most impetus for building controls has been_____.

    a. presidential interest

    b. tragedy

    c. industry push

    d. builders' desire

    Reference _____

4. The question of accepting new or innovative products in a jurisdiction is the responsibility of _____ .

    a. ICC-ES

    b. IAS

    c. the code official

    d. the designer of record

    Reference _____

5. The purpose of IAS is to_____ .

    a. promote the World Trade Organization (WTO)

    b. establish the qualification of testing laboratories and inspection agencies and fabricators

    c. test products for manufacturers

    d. advise design professionals on new construction methods

    Reference _____

6. When the question is asked, "Why do we need building laws?" it would be proper to say that_____ .

    a. lives and property have been lost because of their absence

    b. they keep insurance rates down

    c. Congress requires them

    d. all of the above

    Reference _____

7. Reduction in fire losses over time is thought to have occurred because of_____ .

    a. the invention of concrete and steel

    b. improved inspection, greater fire prevention efforts, and improvements in fire protection systems

    c. the efforts of politicians

    d. none of the above

    Reference _____

8. Local ordinances concerning design issues that are likely to be upheld in court will be based on_____.

    a. aesthetics

    b. homeowners association requirements

    c. architectural drawings

    d. demonstrable public safety

    Reference _____

9. The appeal of tall buildings was enhanced by the development of_____.

    a. curtain wall construction

    b. safe elevators

    c. the invention of helicopters to provide rooftop exiting

    d. all of the above

    Reference _____

10. Building regulation is currently entering an era where building codes are becoming more open to_____.

    a. limiting development

    b. reducing building size

    c. accessibility for the disabled

    d. ecological awareness

    Reference _____

11. Accreditation services provided by IAS provide which of the following important information to the building official?

    a. Testing labs are credible.

    b. Inspection agencies are competent.

    c. Fabricators are doing their job in good quality.

    d. All of the above.

    Reference _____

12. New and innovative building products and components are evaluated by which of the following entities for compliance with the code?

    a. ICC-ES

    b. IAS

    c. ACI (American Concrete Institute)

    d. AISC (American Institute of Steel Construction)

    Reference _____

13. The national program initiated by building officials to focus attention on the importance of codes and code enforcement is called_____.

    a. Fire Prevention Week

    b. Building Code Awareness Week

    c. Building Safety Week

    d. Safe Construction Week

    Reference _____

14. Special inspection agencies are accredited by IAS through assessment of the corporate offices to ensure that the agency has_____.

    a. good customer service

    b. handicap accessible facilities

    c. a diverse workplace

    d. a well-documented quality management system

    Reference _____

15. The Senate Committee of Reconstruction and Production, appointed in 1920, found that the country's building codes were_____.

    a. not legally defensible

    b. not well written

    c. not based on scientific data

    d. not broad enough in scope

    Reference _____

True/false

1. The building official has a responsibility to actively foster awareness of the importance of building control.

   _____

   Reference _____

2. Abatement of objectionable odors, vibrations or noise in a building is recognized as a legitimate subject for the use of police power.

   _____

   Reference _____

3. The claim of the manufacturer or professional engineer of the adequacy of a material is always sufficient to guarantee compliance of the material with the code.

   _____

   Reference _____

4. Building controls have been in effect only since the early 1900s.

   _____

   Reference _____

5. ICC-ES Acceptance Criteria are developed solely by ICC staff.

   _____

   Reference _____

Completion

1. ICC-ES has evaluation reports for _____ and _____.

   Reference _____

2. The first three model code groups were _____, _____ and _____.

   Reference _____

3. The primary intent of building regulation is to provide reasonable controls for the
    _____ and_____.

    Reference _____

4. The types of tragedy that have led to building standards are _____,
    _____ and _____.

    Reference _____

5. Two apparent reasons for inadequately staffed building departments are _____
    and _____.

    Reference _____

Chapter 2
Building Codes and Federal Influences

OBJECTIVE: To learn the purpose of a code and how codes are developed.

LESSON NOTES: Chapter 2 details the ICC code format, the Governmental Consensus Process in code writing, the difference between performance and prescriptive codes, and federal influences on code development.

KEY POINTS:
- What is the general definition of a building code?
- What is the focal point of a building code?
- What is the difference between a building code and a housing code?
- What is the difference between a prescriptive code and a performance code?
- What is the purpose of *International Building Code*® (IBC®) Section 104.11 "Alternate materials, design, and methods of construction and equipment"?
- Why would a jurisdiction adopt the *ICC Performance Code® for Buildings and Facilities*?
- What are the major topics that best describe the format of the IBC?
- How can the building official prepare for unexpected disasters or tragic events?
- What code grew out of the need to address Southern California fires?
- What gives a model code its legal status?
- Why does a code change?
- What is the ICC code change schedule?
- What are the characteristics of the ICC Governmental Consensus Process?
- What two federal government agencies influence the code development process?
- What is the basic responsibility and mission of HUD (Housing and Urban Development)?
- What services does HUD provide toward accomplishing its goals?
- What is the mission of NIST (National Institute of Standards and Technologies)?
- What are the programs within NIST that carry out its mission?
- Who should be involved in the code making process?
- Why is involvement by many groups important to the code making process?

Quiz

Chapter 2

Multiple Choice

1. What is a consideration of the modern building code?

    a. Enforceability

    b. Cost impact of the requirement

    c. Life safety

    d. All of the above

    Reference _____

2. The federal agency that provides mortgage insurance on loans for single-family and multifamily manufactured homes and hospitals is _____.

    a. FHA (Federal Housing Administration)

    b. VA (U.S. Department of Veterans Affairs)

    c. Fanny Mae

    d. Freddy Mac

    Reference _____

3. Ultimately, a _____ decides if the proposed alternative is equivalent to the material or method specified in the prescriptive provisions of the code.

    a. contractor

    b. owner

    c. designer

    d. building official

    Reference _____

4. The federal agency whose mission it is to promote U.S. innovation and industrial competitiveness is _____.

    a. National Institute of Standards and Technologies (NIST)

    b. Department of Energy (DOE)

    c. American Society of Heating, Refrigerating, and Air-Conditioning Engineers (ASHRA)

    d. American National Standards Institute (ANSI)

    Reference _____

5. The current ICC procedure for code revisions occurs over a _____ month period.

    a. 12

    b. 18

    c. 24

    d. 36

    Reference _____

6. New codes are published every _____ years.

    a. 2

    b. 3

    c. 4

    d. 5

    Reference _____

7. The ICC governmental consensus process provides opportunity for _____ to participate in public hearings for code change proposals.

    a. materially-impacted interests

    b. fire officials

    c. architects

    d. all of the above

    Reference _____

8. An issue to be considered in implementing building department procedures to prevent or minimize the effect of disastrous or tragic events include _____.

   a. adopting the latest building codes

   b. coordinating plan review between various departments

   c. implementing high re-inspection fees

   d. providing customer friendly service

   Reference _____

9. The primary reason for FHA standards is to protect _____.

   a. the buyer's property from natural disasters

   b. the developer's financial stability

   c. the government's financial commitment

   d. the building official from liability

   Reference _____

10. The purpose of involving all interested parties in the code development process assures that _____ .

    a. the government's interests are protected

    b. a stable market for sales of the codes is available

    c. the provisions published in the code represent the most up-to-date thinking

    d. no one can complain about what is published

    Reference _____

11. The program under NIST that carries out its mission to develop and disseminate "best practices" for management and operation is _____ .

    a. the Baldridge National Quality Program

    b. the Manufacturing Extension Partnership

    c. the Advanced Technology Program

    d. none of the above

    Reference _____

12. The primary feature of performance codes is the _____ they provide.

    a. stability of structure

    b. various appendices

    c. flexibility

    d. illustrations

    Reference _____

13. A housing code specifies how a residence will be _____.

    a. mortgaged

    b. constructed

    c. zoned

    d. maintained

    Reference _____

14. At the final code hearings only the _____ vote to approve or disapprove the proposed code changes.

    a. ICC Code Change Committee members

    b. public safety officials

    c. ICC staff

    d. super delegates

    Reference _____

15. NIST is an acronym for the _____ .

    a. National Institute for Sprinkler Technology

    b. National Industrial Society of Technology

    c. National Incident and Safety Training

    d. National Institute of Standard and Technology

    Reference _____

True/false

1. NIST is a regulatory agency.

    _____

    Reference _____

2. Public hearings play no role in the code making process.

    _____

    Reference _____

3. The ICC Governmental Consensus Process includes an appeals process.

    _____

    Reference _____

4. A housing code and a residential building code contain the same requirements.
    _____

    Reference _____

5. A model code has legal status once it has been adopted by a legislative body as the law of a community.

    _____

    Reference _____

Completion

1. Companies feel more comfortable working with NIST because it is a
    _____.

    Reference _____

2. Two examples of code provisions that are prescriptive in nature but performance based are: _____ and _____.

    Reference _____

3. An alternate material, design or method of construction shall be approved when the building official finds that it is satisfactory and _____, and that the material, method or work offered is for the purpose intended, and _____.

    Reference _____

4. Name two broad program areas of NIST.

    _____

    _____

    Reference _____

5. Name three imperative characteristics of the code development process.

    _____

    _____

    _____

    Reference _____

3

Chapter 3
The Legacy Model Code Groups, Their Codes and the Roots of the International Code Council

OBJECTIVES: To become familiar with how the model code groups in the U.S. formed, grew, joined and became the foundation for the International Code Council.

LESSON NOTES: This chapter describes how BOCA, ICBO and SBCCI, were created and, along with other organizations such as CABO (Council of American Building Officials), became the building blocks for the ICC. This chapter also describes the goals of the ICC, the make-up of the ICC, how it functions and what the future holds for the model code.

KEY POINTS:
- What did the failure of the rebuilding efforts after the Chicago fire teach about the effectiveness of the political process in code development?
- What was the threat that led to the realization that construction regulation was necessary?
- What was the first model code?
- What new technology in 1892 led to more fires?
- What building code developed from the *National Building Code*?
- What led to the creation of the first building officials' organization in the U.S.?
- What organization took on the task of producing the first code developed by building officials? To what did the organization's name later change?
- What was the name of the second model code developed by building officials, and which organization developed it?
- What was the name of the third model code developed by building officials, and which organization developed it?
- Which states and cities developed and maintained their own codes during the time of the first model codes?
- When was the first edition of the *Uniform Building Code* (UBC) published?
- What was the primary reason for the formation of the Pacific Coast Building Officials Conference?
- What policy was strictly followed during the development of the UBC?
- What U.S. departments adopted the UBC?
- Who were the founding members of SBCCI?

- What was Clement's motivation to develop a model code?
- When was the first edition of the *Southern Standard Building Code* published?
- Which organization was the oldest building officials' association in the U.S. ?
- What was BOCA's motivation for developing a model code? After what did it pattern its procedures?
- What organization formed out of meetings between the officers of the three model code groups?
- What was the purpose of CABO?
- What were the accomplishments of CABO?
- What was the basis for the formation of the BCMC (Board for the Coordination of the Model Codes)?
- What were the responsibilities of the BCMC?
- What organizations made the early case for the development of one model code?
- What was the first successful attempt to coordinate the model codes?
- What were the foreseen benefits of a single model code?
- What were the obstacles to uniting the three model code organizations?
- Why was the name "International Code Council" chosen?
- What was the first code selected for development by the ICC?
- When was the ICC officially formed?
- What were the principles used in the development of the ICC codes?
- When and where was the first joint conference between the three model code groups held?
- When and where was the ICC consolidation finalized?
- Who was the first ICC CEO?
- What are the departments and subsidiaries of the ICC?
- What are the vision, mission and goals of the ICC?

Quiz

Chapter 3

Multiple Choice

1. Which organization led the national effort to eliminate charges of nonuniformity within the model codes?

    a. ICBO

    b. CABO

    c. NFPA (National Fire Protection Association)

    d. BCMC

    Reference _____

2. What "new" technology, in the late 1800s, led to fires in buildings that had previously been considered low fire risks?

    a. Portable heating appliances

    b. Telecommunication

    c. Electricity

    d. Lightening rods

    Reference _____

3. The *National Fire Prevention Code* and *National Building Code* were first developed by which organization?

    a. The National Board of Fire Underwriters

    b. The Pacific Coast Building Officials Conference

    c. The Southern Building Code Congress International

    d. The International Conference of Building Officials

    Reference _____

4. The ICC consolidation was finalized at a ceremony in Las Vegas on what date?

    a. January 21, 2000

    b. January 21, 1999

    c. January 21, 2001

    d. January 21, 2003

    Reference _____

5. The ICC tag line (slogan) is _____.

    a. People Helping People Build a Better Building

    b. People Helping People Through Code Enforcement

    c. People Helping People Build a Safer World

    d. People Helping People Through Construction Regulation

    Reference _____

6. Following the Chicago Fire, the City of Chicago conceded to the National Board of Fire Underwriters that they had failed in their rebuilding efforts because _____.

    a. they had no fire departments

    b. the inspectors were overworked

    c. the mayor passed away

    d. of the threat of losing insurance coverage

    Reference _____

7. The Pacific Coast Building Officials Conference developed the _____.

    a. *Uniform Building Code*

    b. *Standard Building Code*

    c. *National Building Code*

    d. *Basic Building Code*

    Reference _____

8. Prior to formation of the ICC, which code was adopted by the federal Departments of State, Energy, Defense and Agriculture?

   a. *Standard Building Code*

   b. *Uniform Building Code*

   c. *Basic Building Code*

   d. *National Building Code*

   Reference _____

9. The first code selected to be merged by the model code groups was the _____.

   a. *International Building Code*®

   b. *International Fire Code*®

   c. *International Property Maintenance Code*®

   d. *International Plumbing Code*®

   Reference _____

10. What organization issued a white paper making the case for the benefits of a single set of consolidated codes for the nation in 1973?

    a. The National Fire Protection Association (NFPA)

    b. The U.S. Department of Justice

    c. The American Institute of Architects

    d. National Board of Insurance Underwriters

    Reference _____

11. The 1993/1994 editions of all three model codes were produced in what successful coordination effort?

    a. All had the same name for the same code.

    b. All had the same price.

    c. All had the same amendments.

    d. All had the same format.

    Reference _____

12. The first Chief Executive Officer of the ICC was _____.

   a. Rudolph Miller

   b. M.C. Woodruff

   c. Bill Tangye

   d. M.L. Clement

   Reference _____

13. The three previous model code organizations are now known as _____.

   a. the national code organization

   b. the legacy model code organizations

   c. the code official organizations

   d. the real model codes organizations

   Reference _____

14. The National Board of Fire Underwriters' mission in code development was to
   _____.

   a. ensure profits

   b. ensure political involvement

   c. ensure life safety

   d. ensure public awareness

   Reference _____

15. The first edition of this code was distributed free of charge in 1905 to contractors,
   architects, fire marshals and technical schools, and every city with a population over
   5,000.

   a. *International Plumbing Code*®

   b. *Standard Building Code*

   c. *Uniform Building Code*

   d. 1905 *National Building Code*

   Reference _____

True/false

1. The *National Building Code* was completely controlled by the association of insurance underwriters.

   _____

   Reference _____

2. The first ICC Board of Directors had difficulty reaching agreement on the proposed bylaws, structure and transition plan because of block voting by each of the three model code organizations.

   _____

   Reference _____

3. The I-Codes® are used exclusively in the United States.

   _____

   Reference _____

4. Before the *National Building Code*, each municipality wrote its own code, which was often based more on the demands of special interests than on technical merit.

   _____

   Reference _____

5. Building officials had a great deal of influence with the National Board of Fire Underwriters.

   _____

   Reference _____

Completion

1. Name two responsibilities of the National Coordinating Council (NCC) (later named BCMC):

   _____

   _____

   Reference _____

2. Name two goals of the ICC:

   _____

   _____

   Reference _____

3. The organization that undertook the task of producing the first code developed by building officials was _____ , which later became known as _____ .

   Reference _____

4. Name two departments of the ICC:

   _____

   _____

   Reference _____

5. Name two of the principles used in the development of the I-Codes®:

   _____

   _____

   Reference _____

4

Chapter 4
Construction Codes and Standards

OBJECTIVES: To describe the process of creating each of the International Codes® (I-Codes®) and to introduce the agencies that produce standards and/or testing functions that are referenced in the I-Codes

LESSON NOTES: This chapter details the history behind each of the International Codes (I-Codes) and the methodology involved in their development. The standard organizations and testing agencies that are recognized by the ICC are discussed historically; missions are described; and functions are provided.

KEY POINTS:
- What has created the need for a unified system of building codes and regulations?
- What building safety needs that are reflected in the development of the I-Codes do building officials consider important?
- During the consolidation of the three model codes, what was given careful consideration?
- What were the general principles that were followed in the resolution of the technical differences between the legacy model codes?
- How often are the I-Codes published?
- Who can submit a proposed code change to the ICC?
- What is the process for a code change to be adopted?
- When was the first draft of the *International Building Code*® (IBC®) published?
- What is the purpose of the *International Energy Conservation Code*® (IECC®)?
- What is the intent of the *International Existing Building Code*® (IEBC®)?
- Why was the *International Fire Code*® (IFC®) developed?
- What is the intent of the *International Fuel Gas Code*® (IFGC®)?
- When was the first *International Mechanical Code*® (IMC®) published?
- What was the first jointly developed code by the ICC?
- What is the purpose of the *International Private Sewage Disposal Code*® (IPSDC®)?
- What is the purpose of the *International Property Maintenance Code*® (IPMC®)?
- What structures does the *International Residential Code*® (IRC®) regulate?
- Why was the *International Wildland-Urban Interface Code*® (IWUIC®) developed?
- What is the intent of the *International Zoning Code*® (IZC®)?

- What is the intent of the *ICC Performance Code® for Buildings and Facilities*?
- What do standards organizations referenced in the ICC have in common?
- Why was ASTM (Previously American Society for Testing and Materials) originally formed?
- What is the purpose of ASTM International?
- Who is represented on ASTM International committees?
- What is the mission of the International Institute of Standards and Technology (NIST)?
- What is the purpose of the American National Standards Institute (ANSI)?
- What is "Conformity Assessment"?
- What was the original objective of the International Standards Organization (ISO)?
- What U.S. organization is the sole U.S. representative to the ISO?
- What is the aim of the ISO?
- What are the three types of membership in the ISO?
- What technology was the initial target of NFPA attention?
- In what main activities does UL (Underwriters Laboratories) engage?
- What devices engaged the original attention of UL?
- Who founded UL?
- What is the purpose of the American Gas Association (AGA)?
- Who originally formed FM Global?
- What is the focus of FM Global?
- How did the idea of FM Global begin?
- What is the purpose of NSF International?
- What is NSF International most noted for?

Quiz

Chapter 4

Multiple Choice

1. ASTM was originally formed because of concern about _____ .

    a. fires in chimneys in residences

    b. engine fires in cars

    c. concrete failure in tall buildings

    d. frequent rail breaks in the railroad industry

    Reference _____

2. The I-Codes are published every _____.

    a. year

    b. two years

    c. three years

    d. four years

    Reference _____

3. Underwriters Laboratory was founded because of the recognition of the need for examination of all devices that would be _____.

    a. energized by electricity

    b. used in aircraft

    c. used for home heating

    d. imported from other countries

    Reference _____

4. What contribution of FM Global is of primary interest to building and fire code officials?

   a. Standpipe systems

   b. Sprinkler systems

   c. Fire alarm systems

   d. Security systems

   Reference _____

5. What was given careful consideration during the consolidation of the legacy model codes?

   a. Differences

   b. Similarities

   c. Format

   d. Use of regional terminologies

   Reference _____

6. The term used to describe steps taken by both manufacturers and independent third parties to determine fulfillment of standards requirements is _____.

   a. unity assessment

   b. conformity assessment

   c. standard assessment

   d. partnership assessment

   Reference _____

7. Which organization develops safety and performance standards for appliances and accessories fueled by natural gas, LP gas and hydrogen gas?

   a. Associated General Contractors (AGC)

   b. National Fire Protection Association (NFPA)

   c. FM Global

   d. CSA America, Inc.

   Reference _____

8. What was the primary driver for the creation of the ICC?

    a. Cost effective publications

    b. Demand for a single building code

    c. Stronger enforcement authority for building officials

    d. Reduced travel expenses for those involved in the code hearings

    Reference _____

9. Which organization, one of whose purpose is to promote U.S. innovation, is under the U.S. Department of Commerce?

    a. ANSI

    b. NFPA

    c. NIST

    d. ASTM

    Reference _____

10. The first edition of the IBC was published in _____.

    a. 2000

    b. 2001

    c. 2002

    d. 2003

    Reference _____

11. Which organization coordinates U.S. standards with international standards so that American products can be used worldwide?

    a. ASTM

    b. NFPA

    c. NIST

    d. ANSI

    Reference _____

12. What commonality of purpose does each of the standards organizations have?

    a. To promote U.S. products

    b. Safety

    c. Testing

    d. Production of directories

    Reference _____

13. What I-Code was "designed to provide a framework to achieve defined objectives in terms of tolerable levels of damage and magnitude of design events, such as fire and natural hazards"?

    a. *ICC Performance Code for Buildings and Facilities*

    b. IFC

    c. IPMC

    d. IRC

    Reference _____

14. Which organization serves as the spokesperson, clearing house and catalyst for natural gas utilities?

    a. CSA America

    b. AGA

    c. ICC

    d. NSF International

    Reference _____

15. Which organization was created to "facilitate the international coordination and unification of industrial standards"?

    a. NFPA

    b. NIST

    c. ISO (International Standards Organization)

    d. ANSI

    Reference _____

True/false

1. The American National Standards Institute is a for-profit organization that facilitates the development of standards for products, services, processes and systems in the United States.

   _____

   Reference _____

2. Other than ICC staff, anyone may submit a change to an I-Code.

   _____

   Reference _____

3. The *International Plumbing Code* (IPC) was the first jointly developed code by ICC.

   _____

   Reference _____

4. Producers are not allowed to participate on an ASTM committee.

   _____

   Reference _____

5. NFPA's original membership was restricted to those involved in the insurance industry.

   _____

   Reference _____

Completion

1. The three most notable areas of work for the NSF are:

   _____

   _____

   _____

   Reference _____

2. The *International Residential Code* (IRC) regulates_____.

   Reference _____

3. The I-Codes are designed to meet the need for contemporary construction and fire codes, using requirements that emphasize performance through model code regulations that will result in:

_____and

_____.

Reference _____

4. ICC code development committees are of a _____interest.

Reference _____

5. On what intent were the principles of consolidating the legacy model codes based?

_____ and

_____.

Reference _____

Chapter 5
Building Regulations Around the World

OBJECTIVES: To provide an overview of building codes and building regulations from several countries worldwide, as well as the international role of the ICC.

LESSON NOTES: This chapter provides insight into the methods of development, approach and enforcement of codes and building regulations for several countries around the world. The chapter also looks at current trends with regard to the future of those regulations and codes. The ICC Evaluation Service (ICC-ES) and the International Accreditation Service (IAS) interaction at the international level for product and system approval and accreditation is discussed further.

KEY POINTS:
- What is the purpose of a building regulatory system?
- What elements does a building regulatory system typically include?
- At what governmental level do most countries develop building regulations?
- At what level do most countries normally enforce building regulations?
- At what level is Australia's building regulatory system developed?
- What Australian organization was established to pool the country's resources for the development of building regulations?
- What are the cross-continental differences in the Australian building environment?
- Describe the hierarchical format of the *Australian Performance Building Code of Australia* (BCA).
- At what governmental levels can the Australian code be enforced?
- What organization is responsible for code development in Canada?
- What are the model national codes in Canada?
- What are some of the recent changes in the code system in Canada?
- What is the scope of the Canadian codes?
- What are the differences between performance based codes and objective based codes in Canada?
- What is the role of the Canadian Commission on building and fire codes as far as code development is concerned?
- What is the role of the Institute for Research in Construction as far as code development in Canada is concerned?

- What is the role of the Canadian Codes Center for code development?
- Who has jurisdiction over construction in Canada?
- What office has primary responsibility for code development for England and Wales?
- What organization is responsible for building regulation in Japan?
- What is the law that regulates building codes in Japan?
- What is the current state of building regulations in Mexico?
- What organization is working to "generate uniform systems and regulations to harmonize the criteria for quality housing" in Mexico?
- At what level is the building regulatory system written and mandated in New Zealand?
- What countries participate in the Nordic Committee on Building Regulation?
- What is the purpose of the Nordic Committee on Building Regulation?
- What led to changes in building regulations in Norway in the mid-1990s?
- When was the first code in Spain approved?
- Why was there no prior code in Spain?
- What organization develops Swedish codes?
- What is the Swedish code approach?
- What are the methods that may be followed to obtain an ICC-ES evaluation report for a global product?
- What is the difference between U.S. accreditation and accreditation in other countries?
- What is the benefit for U.S. manufacturers in utilizing IAS accredited laboratories?
- What is the purpose of IAS accredited calibration labs?
- What is the purpose of IAS accredited testing labs?

Quiz

Chapter 5

Multiple Choice

1. What organization is responsible for building regulation in Japan?

    a. The Ministry of Buildings and Construction

    b. The Ministry of Building Regulations

    c. The Ministry of Land, Infrastructure and Transport

    d. The Ministry of Code Development

    Reference _____

2. What organization has entered into an agreement with Mexico to assist in the development of their *Residential Building Code*?

    a. Home Builders Association

    b. ICC

    c. U.S. Department of Housing and Urban Development (HUD)

    d. Council of American Building Officials (CABO)

    Reference _____

3. What is the provision in Spain's NBE (Normas Basicas de la Edificacion—which means Basic Building Standards) that allows adoption of solutions differing from the building regulations, provided they still fulfill the objectives of the regulation?

    a. Alternate Methods and Materials

    b. Performance Clause

    c. Objective Clause

    d. Freedom Clause

    Reference _____

4. Which country claims to have developed the world's first objectives-based code?

   a. United States

   b. Canada

   c. Sweden

   d. Japan

   Reference _____

5. How many model codes are developed and updated by the Canadian Commission of Building and Fire Codes?

   a. 6

   b. 9

   c. 12

   d. 4

   Reference _____

6. The countries of Denmark, Finland, Iceland, Norway and Sweden have formed what organization to be responsible for building regulations?

   a. Nordic Committee on Building Regulations

   b. Nordic Committee for Code Development

   c. Nordic Committee for Construction and Rehabilitation

   d. Nordic Committee for Building Safety

   Reference _____

7. The organization created in Australia to develop a nationally-consistent, performance-based building regulatory system that was efficient, cost effective and met community, industry and national needs was the _____.

   a. Australian Department of Codes and Standards

   b. Australian Board of Architecture and Engineering

   c. Australian Ministry of Construction

   d. Australian Building Codes Board

   Reference _____

8. The Canadian Code Centre technical advisors are made up of _____.

    a. architects and engineers

    b. engineers only

    c. fire service professional and engineers

    d. architects only

    Reference _____

9. In the early 1990s, Australia's Building Regulatory Review Task force reported that the country's building regulatory systems were causing cost overages to industry, government and the community of up to how much each year?

    a. $1 million

    b. $100 million

    c. $300 million

    d. $1 billion

    Reference _____

10. Documentation and certification of the qualifications and competence of designers and contractors by the building authorities was one of the most important changes in the Planning and Building Law for what country?

    a. Japan

    b. Canada

    c. Norway

    d. Mexico

    Reference _____

11. Accreditation by the IAS results in what added benefit to U.S. manufacturers?

    a. lower export taxes

    b. federal subsidies

    c. free advertisement

    d. greater international opportunities

    Reference _____

12. In an attempt to increase flexibility in design and produce a more intelligent system, England's Buildings and Regulations, 1985 publication, was reduced from 307 pages to

_____.

    a. 250 pages

    b. 23 pages

    c. 129 pages

    d. 76 pages

    Reference _____

13. The code in what country contains three levels: Objectives, Functional Requirements and Performance Criteria for each clause?

    a. Wales

    b. Japan

    c. New Zealand

    d. Canada

    Reference _____

14. The Swedish Building Regulation has what embodied within it?

    a. Design criteria

    b. Alternate methods and materials

    c. Commentary

    d. Both a and b

    Reference _____

15. A 1994 report in Norway that prompted the government to amend the Planning and Building Law concluded that it was likely that what percent of the total output of the construction industry in Norway was used to remedy building defects?

    a. 25 percent

    b. 15 percent

    c. 10 percent

    d. 5 percent

    Reference _____

True/false

1. Codes provided through the Building Standards Law in Japan are uniform across the country.

   _____

   Reference _____

2. IAS is the oldest accreditation body in North America to have earned international recognition and acceptance for all of its accredited laboratories and inspection agencies.

   _____

   Reference _____

3. The Canadian Commission on Buildings and Fire Codes publishes all of Canada's building regulatory codes.

   _____

   Reference _____

4. Canadian fire codes do not contain retroactive requirements but allow buildings to remain compliant with the codes under which they were built in all respects.

   _____

   Reference _____

5. In England and Wales, the enforcement of the building regulations is undertaken at the local level and can be undertaken by public or private services.

   _____

   Reference _____

Completion

1. What are the three parts of the Building Standards Law in Japan?

   _____

   _____

   _____

   Reference _____

2. What are the three elements typically found in a building regulatory system?

    _____

    _____

    _____

    Reference _____

3. What are the four tiers of Australia's Performance BCA (*Building Code of Australia*)?

    _____

    _____

    _____

    _____

    Reference _____

4. What are the three options for code users in Japan, under the current performance based code system?

    _____

    _____

    _____

    Reference _____

5. Name two of significant cross-continental differences in the Australian building environment?

    _____

    _____

    Reference _____

Chapter 6
The Department

OBJECTIVES: To develop a thorough understanding of the organizational structure and the challenges encountered by the building department.

LESSON NOTES: This chapter answers questions that frequently arise concerning how to best organize a building department, the desirability of autonomy of the building department, the necessity of uniform identification for jobs and titles in the building department, and the importance of recognition of the building department in the municipal government. In addition, challenges that affect small and large building departments are analyzed. The benefit of accreditation of building departments is discussed.

KEY POINTS:
- What is the mission of the building department?
- What key factors affect the organization of a building department?
- What are the primary duties of the building department?
- How does a building department accomplish its duties?
- What factors can influence the staffing of a building department?
- How can having different codes in different jurisdictions affect administering the code in any one jurisdiction?
- What type of technological advances caused change in the construction and design industry?
- How does the adoption of one common set of codes across the nation benefit the building department?
- How does creating a board of appeals help the building official?
- What are the merits and problems associated with separating the building department from the public works department?
- What role do performance measures play in managing a building department?
- What are the intangible qualities that are important to a building department?
- Ideally, how should fees be determined?
- What can cause a building department to show a profit?
- What are some nonrevenue-producing activities that can take up quite a bit of time in the building department?
- What is the objection to the fixed fees concept?

- What is one way to reduce the effect that poor quality submissions have on the building department?
- What types of additional fees might be considered so as to reduce the number of no-fee functions?
- What is an Enterprise Fund, and how is it useful?
- Why are city and county managers opposed to an enterprise fund?
- What must be considered and determined when issuing a refund of fees?
- What are the impacts of lack of uniformity in building department job titles?
- What duty in the building department is most important and why?
- How is plan checking accuracy important to the inspector?
- What are the basic tasks of the inspector?
- To whom is the building department's highest obligation?
- What does the fact that the code is a minimum mean to the enforcement official?
- Who are special inspectors?
- What are the required special inspections in the ICC?
- Whose responsibility is it to determine the qualifications of the special inspector?
- What organization has an accreditation program for the special inspector?
- What is the function of the Insurance Services Office (ISO)?
- What is the intent of the Building Codes Effectiveness Grading Schedule (BCEGS) program?
- How do building departments benefit from IAS Building Department Accreditation?
- What is IAS On-Site Evaluation?
- What are some of the problems affecting small jurisdictions?
- How are staffing issues different for small jurisdictions?
- What are the special needs of large jurisdictions?
- What are some of the ways that large jurisdictions can address their special needs?
- Why is underrepresentation at code hearings a problem for large jurisdictions?

Quiz

Chapter 6

Multiple Choice

1. Which entity usually issues a certificate of compliance that is required by some states to accompany the sale of existing buildings?

    a. Municipal finance department

    b. Building department

    c. State department of licensing

    d. State fire marshal

    Reference _____

2. The organization that has an accreditation program for special inspectors is _____.

    a. Insurance Services Office (ISO)

    b. International Accreditation Service (IAS)

    c. ICC Evaluation Service (ICC-ES)

    d. CBS

    Reference _____

3. Among other municipal departments, which seems to be most closely aligned with the needs of a small building department?

    a. The parks and recreation department

    b. The police department

    c. The fire department

    d. The finance department

    Reference _____

4. The final authority for determining the qualifications of the special inspector is the responsibility of _____.

   a. IAS

   b. the designer of record

   c. the Director of Building and Safety

   d. the building official having jurisdiction

   Reference _____

5. An area of the building department services that is not always given the attention it needs is _____.

   a. plan checking

   b. inspection

   c. fee collection

   d. training

   Reference _____

6. By law, the municipality has an obligation to _____.

   a. property owners

   b. tax payers

   c. the elected officials

   d. the public at large

   Reference _____

7. The mission of the building department includes _____.

   a. protecting the lives and safety of residents

   b. preserving the quality of life

   c. contributing to the economic development of the jurisdiction

   d. all of the above

   Reference _____

8. The most influential factor in determining the size of the building department is
   _____.

   a. age of the buildings in the jurisdiction

   b. industrial growth in the jurisdiction

   c. size of the municipality

   d. whim of the elected officials

   Reference _____

9. Some of the personal responsibility of building officials for code interpretations can be relieved by the creation of _____.

   a. a website containing jurisdictional code interpretation

   b. a link to the ICC website

   c. commentary to the codes

   d. a board of appeals

   Reference _____

10. What tool is essential to the building department manager to improve efficiency and improvement of all functions in the building department?

    a. Performance measures

    b. Competent clerical assistance

    c. Management training

    d. IAS accreditation

    Reference _____

11. Assessing fixed fees for plan checking and inspection regardless of the project type and complexity is _____.

    a. most simple

    b. most direct

    c. most unfair

    d. all of the above

    Reference _____

12. What organization serves the insurance marketplace with statistical, actuarial and underwriting information?

    a. ISO

    b. IAS

    c. FM Global

    d. UL

    Reference _____

13. The Building Code Effectiveness Grading Schedule (BCEGS) program is designed to assess the building codes in effect in a particular community and how the community enforces the codes with special emphasis on _____.

    a. consistency of interpretation

    b. quality of inspections

    c. mitigation of losses from natural hazards

    d. building department organization

    Reference _____

14. The strength of the IAS accreditation program is that it gives the jurisdiction _____.

    a. a certificate for the public to see

    b. a list of problems it must address

    c. better insurance rates

    d. a comprehensive business process standard

    Reference _____

15. The phrase "You can't manage what you can't _____" is appropriate when discussing the efficiency of a building department.

    a. see

    b. fund

    c. measure

    d. document

    Reference _____

True/false

1. To rely on an inspector to track down design errors is realistic.

    _____

    Reference _____

2. City and county managers are usually opposed to enterprise funds because they place restrictions on their authority in the budget process.

    _____

    Reference _____

3. It is acceptable and the best practice for an inspector to act on any change to the approved plans that has not been authorized through the plan check process.

    _____

    Reference _____

4. A building inspector should maintain good public relations by being a "facilitator" rather than being only a "regulator."

    _____

    Reference_____

5. A building code is essentially a remedial law, intended to be construed liberally and beneficially for safety.

    _____

    Reference _____

Completion

1. Name two disciplines in which special inspectors can be certified.

    _____

    _____

    Reference _____

2. What three services constitute the basic foundation on which the building department is structured to fulfill its duties?

_____

_____

_____

Reference _____

3. What are three intangible qualities that pose a challenge to building department management?

_____

_____

_____

Reference _____

4. Generally, the objection to the fixed fee concept is that it_____.

Reference _____

5. Name three characteristics of a jurisdiction and its building department that the IAS accreditation process for a building department will require information on.

_____

_____

_____

Reference _____

7

Chapter 7
The Development Permit Process

OBJECTIVES: To learn how a building department can best protect the public while advancing the community's growth and development.

LESSON NOTES: Chapter 7 describes the various elements of the development process from a building department aspect, their purposes, their influence over the development process, and best practices in the process.

KEY POINTS:
- What are the conflicting forces in the role of the building department?
- What is the primary purpose of the development process in any jurisdiction?
- What are the reasons for a jurisdiction's oversight of development?
- What are the three broad functional areas of the development process most often found in the U.S.?
- What is the basis for the typical zoning process?
- What is the purpose of the subdivision development process?
- What are the typical functions of a building permit process?
- What can help eliminate regulatory confusion within a jurisdictional department?
- What is the purpose of the typical planning department in the development process?
- What is the purpose of the typical public works department in the development process?
- What is the purpose of the typical engineering department in the development process?
- What is the role of the building department in the development process?
- What is the most vital function within the building department in the development process?
- What organization has the most to gain in the development process?
- What is the potential gain of the economic development department in the development process?
- What are the responsibilities of the fire prevention department?
- How can architects and engineers provide assistance to the building department?
- What is the role of the planning and zoning commission or county commissions in the development process?

- Why is an appeals board needed?
- What is the purpose of the zoning board of appeals?
- What are the two broad trends that affect how each jurisdiction designs its development review and inspection process?
- How has the move toward environmental regulations influenced the development process?
- What are the four critical areas for application of best practices to streamline the permitting process?
- What are some examples of regulatory and legal influences that can affect the overall development process?
- What are the best practices of project management in the building department?
- What is the best practice for communication with customers?
- How does the Fairfax County, VA Expedited Building Plan Review Program work? What is the key to its success? What are the benefits? What are its potential drawbacks?
- What is the purpose of the permit ombudsman in the permit process in Clark County, WA?
- What are the features of the One Stop Shop in Carrolton, TX? What are the benefits of this process?
- How has placing the permit application submittal process in a single department benefitted the development process in Portage, MI?
- What areas were deemed priorities for the improvement of the San Diego, CA development process?
- What role has the use of the Web by building departments played in the improvement of some of the case studies?

Quiz

Chapter 7

Multiple Choice

1. In general, which organization provides a formal process for citizen participation and comment on various land-use decisions?

    a. City Council

    b. County Board

    c. Planning and Zoning Commission

    d. Appearance Commission

    Reference _____

2. What process within the municipality typically involves the review and inspection of public infrastructure?

    a. Subdivision

    b. Economic development

    c. Code enforcement

    d. Planning and zoning

    Reference _____

3. What organization in a municipality is held responsible for the administration and enforcement of all regulations and requirements pertaining to site development and construction of buildings?

    a. Engineering

    b. Office of the City Manager

    c. Development Services

    d. Planning and Zoning

    Reference _____

4. The body that holds a hearing to determine the validity of the building department's imposition of rules and regulations is called the _____.

    a. City Council

    b. Board of County Commissioners

    c. Board of Appeals

    d. District Attorney's Office

    Reference _____

5. What department in the jurisdiction is typically responsible for information related to the predevelopment phase?

    a. Economic Development

    b. Permit Department

    c. Utilities Department

    d. Planning Department

    Reference _____

6. The department with responsibility for providing public services such as water distribution, wastewater collection, pavement management, traffic services related to street construction, and storm water management is

    a. Public Works

    b. Engineering

    c. Health Services

    d. Risk Management

    Reference _____

7. This entity's purpose is to enhance the tax base and the business viability of the community.

    a. City Council

    b. Economic Development Department

    c. Office of the City Manager

    d. Downtown Revitalization Committee

    Reference _____

8. This department's main focus is to implement contracts for the design and construction of the jurisdiction's infrastructure.

    a. Engineering

    b. Purchase and Contract

    c. Budget and Analysis

    d. Public Utilities

    Reference _____

9. The typical zoning process addresses requirements based on _____.

    a. a desire for balance between urban and rural uses

    b. the needs of the area's largest employer

    c. population wealth distribution

    d. an adopted comprehensive land use plan

    Reference _____

10. The primary purpose of this department is to regulate and provide reasonable controls for the construction, use and occupancy of buildings and sites as mandated by the jurisdiction.

    a. Building Inspection

    b. Engineering

    c. Facility Services

    d. Risk Management

    Reference _____

11. Many building safety officials believe this function is the most vital part of the development process.

    a. Permit fees

    b. Planning and zoning

    c. Plan review

    d. Subdivision review

    Reference _____

12. Which of the following can provide valuable assistance to the building department to identify flaws in the review and inspection process?

    a. Architects and engineers

    b. Home owners

    c. Outside tax auditors

    d. Government officials

    Reference _____

13. Which of the following is established to impose reasonable conditions in ensuring compliance and protection of property? They may make exceptions to ordinances or other administrative decisions to render justice and equity to the general public for the development of their property.

    a. City Council

    b. County Commissioners

    c. Building inspectors

    d. Zoning Board of Adjustment

    Reference _____

14. Which of the following is most commonly responsible for reviewing the zoning request application to verify that they meet the general development requirements before the application is placed through the formal hearings?

    a. Development Review Committee

    b. Assistant City Manager

    c. City Council

    d. County Commissioners

    Reference _____

15. The building permit process usually resides in the _____ .

    a. Engineering Department

    b. Public Utilities Department

    c. Building Department

    d. Planning and Zoning Department

    Reference _____

True/false

1. There are many ways to manage a government's development process; one size does not fit all.

    _____

    Reference _____

2. It is not necessary for the fire prevention department to be involved in the development process.

    _____

    Reference _____

3. Architects and engineers should be provided with an avenue to meet with staff to resolve design issues that conflict with adopted codes and ordinances.

    _____

    Reference _____

4. The duties of the planning departments are typically performed in two functional areas: current planning and long range planning.

    _____

    Reference _____

5. It is not necessary for the contractor to understand any jurisdictional departments other than the building department.

    _____

    Reference _____

Completion

1. In a jurisdiction, the development process exists for one purpose. It is

    _____

    Reference _____

2. The three broad functions of the development process commonly found in the U.S. are:

    _____

    _____

    _____

    Reference _____

3. The two broad trends that affect how each municipality designs its ultimate development review and inspection process are:

    _____

    _____

    Reference _____

4. The conflicting forces affecting agencies charged with the welfare of the public are:

    _____

    _____

    Reference _____

5. Name two of the four critical areas for the application of best practices for streamlining the permitting process that were identified in the 2004 study of nine jurisdictions.

    _____

    _____

    Reference _____

Chapter 8
The Building Official

OBJECTIVES: To gain a deeper understanding of the role of the building officials and their authority in the Building Department.

LESSON NOTES: This chapter provides explanation of the qualifications, duties, responsibilities and characteristics of the successful building official. Additionally, this chapter provides suggestions for handling situations that arise in the profession. Finally, the building official's code of ethics is discussed in detail

Key Points:
- What is the primary responsibility of the building official?
- What assumptions do owners make when receiving a Certificate of Occupancy?
- What is an Ad Valorem tax system?
- Why is it necessary for a building official to educate elected leadership on the importance of building safety?
- What are some examples of ways to educate the public on building safety?
- How should building officials handle situations in which they are asked for favors?
- How can a building official affect employee morale?
- Why are employee work standards important?
- Why should a building official be concerned with the employee's salary instead of leaving it to the personnel department?
- Why is there a provision in the IBC for modifications, and how is it intended to be used by the building official?
- Based on the duties of the building official, what are some of the qualifications that would be needed to fill that position?
- What is the difference between the formal and informal structures of an organization?
- How can personal relationships impact the structure of an organization?
- How has the role of a manager/leader changed in the last 20 years?
- What are the four steps involved in management?
- With respect to recruitment, what is one of the most important lessons for a manager to learn?
- What are the five basic rules of conduct of a successful building official? How can each one bring success to the department?

- What is the building official's role with respect to discipline and morale of the staff?
- What is the most effective way to approach problem solving?
- How must a manager approach complex problems?
- What is the building official's first step toward delegation?
- How is delegation beneficial to the department?
- Why is it important to develop employees' potential?
- Why is effective communication important within the department?
- What is the difference between formal and informal communication?
- What is the basic premise of the ICC Code of Ethics?
- What is necessary to achieve harmony in the building department? How does the building official affect that?
- Why is it necessary for the building official to subdivide the staff into smaller units?
- Why is it necessary for a building official to be aware of and protect the personal interests of the staff?
- Why is personal appearance important?
- What are some effective ways to deal with the media?
- Why is it important for the building official to develop a relationship with other governmental agencies?
- Why is proper understanding and use of legal terms important to the building official?

Quiz

Chapter 8

Multiple Choice

1. Leadership differs from managing and directing in that it is more motivational and inspirational rather than _____.

    a. technical

    b. domineering

    c. disinterested

    d. political

    Reference _____

2. Most building code enforcement organizations are no longer bureaucracies but operate as _____.

    a. teams

    b. regulators

    c. facilitators

    d. fee collectors

    Reference _____

3. A building official's first step toward delegation is _____ .

    a. deciding to whom to assign the task

    b. asking others for opinions on what to delegate

    c. admitting they do not know everything

    d. defining the assignment

    Reference _____

4. Whenever work is found to be contrary to the provisions of the code, depending on the severity of the violation, the building official may _____.

    a. call the police

    b. call the fire department

    c. fence off the work area to prevent further access

    d. issue a stop work order

    Reference _____

5. To be regarded as modern, the profession that regulates safety has a responsibility to embrace _____.

    a. new technology

    b. complete record keeping

    c. reliance on professional registration of designers

    d. only tried and true methods

    Reference _____

6. When addressing a legislative body it is better for the building official to _____ a prepared statement that is also distributed to the legislators beforehand.

    a. read directly from

    b. prepare a power point of

    c. extemporize from

    d. refer to

    Reference _____

7. The building official is primarily responsible for the _____ associated with buildings or structures.

    a. economic impact on the community

    b. zoning impact

    c. water and sewer impact

    d. human safety

    Reference _____

8. In the quest to make an informed decision, the place to start is with _____.

    a. the spirit and intent of the code

    b. asking opinions of several other code officials

    c. referring to related sites on the internet

    d. all of the above

    Reference _____

9. In the past 20 years, management has evolved such that one of its important characteristics, "directing", has become _____.

    a. coordinating

    b. leading

    c. advising

    d. researching

    Reference _____

10. Many building officials clarify certain commonly asked questions in the form of _____ that are intended to promote uniformity of enforcement by the building code staff.

    a. policy statements

    b. FAQs

    c. newsletters

    d. bulletins

    Reference _____

11. The building official is the appointed officer primarily responsible for the administration and enforcement of the various codes and ordinances related to _____.

    a. economical construction

    b. the jurisdiction's future plans for growth

    c. building safety

    d. homeland security

    Reference _____

12. The key to solving the dilemma of industries having little or no regard for local laws or regulations concerning building safety is/are _____.

    a. vigilant inspections

    b. a moratorium on permits for them

    c. charging more for their permits

    d. education on the importance of building regulation

    Reference _____

13. In the ICC's Code of Ethics, the ICC Certified Individual shall "Place the _____ above all other interests and recognize that the chief function of government is to serve the best interest of all people.

    a. community's economic interest

    b. public's welfare

    c. elected leadership's interest

    d. none of the above

    Reference _____

14. A building official must keep an open mind with respect to the use of new and innovative techniques that may not meet the prescriptive requirements of the law so as to see how the proposal meets the _____ of the law.

    a. intent

    b. letter

    c. performance requirements

    d. none of the above

    Reference _____

15. The authority of the building official is limited to enforcing and interpreting the code and is not allowed to _____ the code.

    a. teach

    b. sell

    c. grant a variance from

    d. all of the above

    Reference _____

True/false

1. The building official has a duty to maintain discipline and morale at the same time.

    _____

    Reference _____

2. It is not the responsibility of the building official to maintain accurate records.

    _____

    Reference _____

3. The building official must never assume responsibility for the mistakes of the staff.

    _____

    Reference _____

4. Because they have trained staff to enforce the codes, it is not important for building officials to know much about the codes .

    _____

    Reference _____

5. The building official should promote the personal interests of all staff at all times.

    _____

    Reference _____

Completion

1. Name two of the four steps defining the leadership role of the building official.

    _____

    _____

    Reference _____

2. Name two of the four steps involved in writing a policy.

    _____

    _____

    Reference _____

3. Name two of the four reasons for one-on-one communication with staff.

    _____

    _____

    Reference _____

4. Three steps toward effective interaction with the media are:

    _____

    _____

    _____

    Reference _____

5. Name two of the general rules of conduct for a building official.

    _____

    _____

    Reference _____

Chapter 9
The Effective Manager

OBJECTIVES: To provide a profile of an effective government manager and the means to become an effective manager.

LESSON NOTES: This chapter illustrates several theories on how a public manager can be effective, including ways to motivate staff, encourage excellence and assist design professionals with adherence to codes for successful projects.

KEY POINTS:
- What is the difference between managing a government agency and managing in the private industry?
- What is the most effective way to gain buy-in from staff and reach goals?
- What is the most important thing for a manager to keep in mind?
- Why is it important for a manager to stay abreast of changes?
- Why is it important for a manager to keep the staff engaged in the goals and objectives of the agency?
- Why must an effective manager be committed to his or her community?
- Why must an effective manager understand the effects of conflict on the community?
- Why is it important for management to keep staff in tune with the code change process?
- Why is it important for the building official to maintain a good reputation with respect to his or her duties?
- How is a leader different from a manager?
- Why must a manager understand the intricacies of programs and functions of his or her staff?
- Why is it important for a manager to gain the trust of his or her staff?
- What actions are necessary when corrective actions must be taken with an employee?
- Why is commitment to the goals of the organization a primary requirement for managers?
- What is involved in the manager's promotion of the future of the community, and why is it important?
- Why is it important for a manager to manage his or her life, both professionally and personally?

Quiz

Chapter 9

Multiple Choice

1. An effective manager shows his or her staff and the community a commitment by_____.

    a. putting in extra hours at work

    b. running for elected office

    c. participating in community activities

    d. promoting rigid enforcement of the laws of the community

    Reference _____

2. When considering steps taken to address a community's hazards, _____ is not an admission of a community's weakness, but rather of its strength and interest in reducing its long-term liability.

    a. a strong code

    b. political involvement

    c. media interest

    d. all of the above

    Reference _____

3. The manager of an organization's leadership is measured by how he or she
   _____.

   a. sticks to the department's budget

   b. relates to the elected officials

   c. motivates staff

   d. none of the above

   Reference _____

4. According to Ralph Shrader, "Leaders are not self created", but are "_____."

   a. born to be leaders

   b. acknowledged by others

   c. appointed to their position

   d. taught to be leaders

   Reference _____

5. Managing without the support of the staff assigned to carry out the overall
   administration and day-to-day activities of the department will _____.

   a. mean it will be necessary to replace the staff

   b. lead to dysfunctional programs

   c. most likely lead to the manager's dismissal

   d. require more staff meetings

   Reference _____

6. Demonstrating leadership through _____ is the most effective way to gain
   buy-in from the staff.

   a. holding staff meetings

   b. delegating work

   c. example

   d. none of the above

   Reference _____

7. Government managers must always keep in mind that they are, first, _____.

   a. a taxpayer, too

   b. a supervisor

   c. an authority on the responsibilities of the department they manage

   d. a caretaker of the public trust

   Reference _____

8. Managers must familiarize their staff with potential_____, which will assist them in the daily application of code provisions.

   a. areas of conflict

   b. interpretations

   c. political favors

   d. media inquiries

   Reference _____

9. An effective manager must balance the needs of the staff with the needs of the _____to carry out the principles of the agency and government leaders.

   a. elected officials

   b. program

   c. budget

   d. citizens

   Reference _____

10. The ten Natural Laws of Successful Time and Life Management by Hyrum W. Smith, CEO of Franklin Quest Company, include _____.

    a. control your life by controlling your time

    b. give more and you have more

    c. to reach any significant goal, you must leave your comfort zone

    d. all of the above

    Reference _____

11. An effective manager who provides feedback and sets goals for the staff encourages them to _____.

    a. excel

    b. compete with each other

    c. complain about the workload

    d. make the position a career

    Reference _____

12. As changes occur in a community, the jurisdiction's code agency must keep the policymakers aware of the goals of _____.

    a. the revenue department

    b. public safety

    c. salary increases for staff

    d. none of the above

    Reference _____

13. Alternative funding mechanisms for the budget of a government agency are _____.

    a. enterprise funding

    b. dedicated funds

    c. special assessment funds

    d. all of the above

    Reference _____

14. Most core values in the new paradigm of government agencies include what as the cornerstone of success?

    a. Excellence

    b. Teamwork

    c. Inclusion

    d. All of the above

    Reference _____

15. Identifying potential sources of conflict in the processes, procedures and desires of the customers and the building department requirements can lead to _____.

a.   more conflicts

b.   tension among staff

c.   creative solutions

d.   none of the above

   Reference _____

True/false

1.   According to the FEMA report, "Promoting the Adoption and Enforcement of Seismic Building Codes," building codes have hurt the economies of the states that have them.

   _____

   Reference _____

2.   A manager does not necessarily need the trust of his or her employees to effectively lead them.

   _____

   Reference _____

3.   An effective manager has no need to consider the future of the community.

   _____

   Reference _____

4.   Building officials have no duty to assist design professionals in understanding that codes are a necessary element in all aspects of the built environment.

   _____

   Reference _____

5.   An effective manager is committed to the organization's goals.

   _____

   Reference _____

Completion

1. Name two issues the specifics of which the manager must understand to effectively manage a government agency.

    _____

    _____

    Reference _____

2. Name two issues that the manager must keep abreast of as the basis for the effective management of a government agency.

    _____

    _____

    Reference _____

3. Name two problems that can possibly increase the cost of construction, cause immediate issues or long-term building problems.

    _____

    _____

    Reference _____

4. Name three principles of leadership defined by Colin Powell, former U.S. Secretary of State.

    _____

    _____

    _____

    Reference _____

5. Name two things that can be improved by including subordinates in the development processes and recognition and award programs.

    _____

    _____

    Reference _____

Chapter 10
Supervision and Training

OBJECTIVES: To discuss and provide an overview of the elements of effective supervision and training.

LESSON NOTES: Chapter 10 provides a description of the important elements of effective supervision. Supervisor motivation of staff, as well as addressing staff needs and what is involved in an effective performance review are discussed. The importance of employee training is emphasized. The necessity and structure of a department procedures manual is provided.

KEY POINTS:
- What is "esprit de corps?"
- What are the principle elements of supervision?
- Why is courage an important characteristic of supervision?
- Why is humility an important trait?
- How does the attitude of the director of the building department affect the attitudes of the subordinates?
- What is not part of the definition of a good leader?
- What are the characteristics most essential to effective administration of a building department?
- How can a supervisor create the "will to work"?
- Do employees' attitude toward their company affect their productivity?
- Why should a supervisor be specific about the goals and tasks to be accomplished?
- What are some of the basic employee needs in the workplace?
- Why is it important for the supervisor to keep up with employee needs and respond to them?
- Why is information exchange an employee need?
- What characteristics of a supervisor give an employee security?
- Why is a performance review beneficial?
- What are the rules for conducting a performance review?
- When an employee needs correction, what are the necessary steps?
- Why is feedback from the supervisor to the employee important?
- What is the importance of manager-employee interaction?

- Why should the supervisor be trained to be a supervisor?
- What is the new concept of performance reviews?
- How does the new concept of performance reviews work and what does it provide to the supervisor and employee?
- Why is a written procedures manual necessary?
- How should the written procedures manual be structured?
- Why should the department procedures manual be provided to the trades?
- Why is extensive training necessary in code enforcement?
- What kind of training is needed in a building safety department?
- Who is responsible for department staff training?
- Why are certification programs beneficial to the code enforcement agency?
- What are the three types of motivation?
- Besides code enforcement inspection certification, what other types of certification are available from the ICC?
- What is special about "special inspector?"

Quiz

Chapter 10

Multiple Choice

1. In order to have assigned tasks completed with competence and consistency it is best to
    _____.

    a. be firm with employees

    b. spell out precise instructions in department procedures manual

    c. hold more staff meetings

    d. hold staff retreats

    Reference _____

2. When an employee is promoted to a supervisory role, both initial and extended training on _____ should be provided, rather than training in the work itself.

    a. supervision

    b. public information

    c. communications skills

    d. managing a budget

    Reference _____

3. The complexities of good supervision are based on _____.

    a. workplace politics

    b. means to award raises fairly

    c. outside influences on workload

    d. differences in people

    Reference _____

4. _____ involves having good moral principles, being upright, honest and sincere.

   a. Having integrity

   b. Good media relations

   c. Working with local elected leaders

   d. Dealing with the public

   Reference _____

5. The building department manager must not only have good technical skills, but also good_____ to influence others.

   a. media relations

   b. management and leadership skills

   c. political connections.

   d. none of the above

   Reference _____

6. Psychologists' term for "the will to work is" _____.

   a. work ethic

   b. integrity

   c. motivation

   d. all of the above

   Reference _____

7. Poor morale is preceded by an employee's uncertainty about _____.

   a. how management views their performance

   b. the security of their 401k

   c. how they compare to their peers in the workplace

   d. who will be their supervisor

   Reference _____

8. _____is the only way that a building department can keep pace with changing and complex construction methods and materials.

    a. The internet

    b. Updating referenced standards available to the staff

    c. Personnel training

    d. Using outside consultants more regularly

    Reference _____

9. Even new employees with certifications and skills must be provided with _____ so that the supervisor can determine their competence.

    a. code books

    b. periodic testing on codes throughout the probation period

    c. oversight

    d. none of the above

    Reference _____

10. _____ are highly skilled persons who are certified and employed to evaluate such critical work as welding, high-strength bolting and other similar specialized areas of construction on the job site.

    a. Registered engineers

    b. Special inspectors

    c. CBOs

    d. Construction managers

    Reference _____

11. The building department should provide training on _____ in addition to training related to technical competency.

    a. credibility

    b. communication

    c. ethics

    d. all of the above

    Reference _____

12. Informal training is important in staff development, but it must accompany a consistent, measurable_____.

    a. training program

    b. work load

    c. productivity data base

    d. none of the above

    Reference _____

13. The effective organization is a result of intentional and appropriate _____.

    a. training and supervision

    b. recruitment

    c. benefits

    d. feng shui

    Reference _____

14. To encourage an employee and reduce job dissatisfaction, better supervisors usually specify the goal or task to be done and give the employee _____.

    a. every detail on how it is to be accomplished

    b. no time limit

    c. some leeway in the way to accomplish it

    d. all of the above

    Reference _____

15. The purpose of a performance review is _____, and not a time to reprimand or humiliate the employee.

    a. just a formality to satisfy the human resources office

    b. the tool by which salary increases are determined

    c. to force the supervisor to talk to the employee

    d. purely constructive

    Reference _____

True/false

1. Although appreciated, commending good work of the employees is not necessary to be an effective leader.

   _____

   Reference _____

2. A building department must always apply the code consistently.

   _____

   Reference _____

3. The cost of not providing adequate employee training may be more that the cost of providing it.

   _____

   Reference _____

4. Studies have shown that an employee's attitude toward their company has little effect on their productivity.

   _____

   Reference _____

5. It is not necessary to share the department's procedure manual with anyone outside of the department.

   _____

   Reference _____

Completion

1. Name two flavors of motivation.

   _____

   _____

   Reference_____

2. Integrity means keeping promises made to employees or, if that is not possible, giving_____.

    Reference_____

3. Name two things describing the best attitude for the building department manager:

    _____

    _____

    Reference_____

4. Effectiveness of instilling the will to work in employees depends on how well managers understand _____.

    Reference_____

5. The need for _____is an additional reason that people work besides financial income.

    Reference_____

Chapter 11
Department Staffing Requirements

OBJECTIVES: To gain a thorough understanding of the various issues associated with providing adequate staffing to perform the services required of the building department.

LESSON NOTES: Examples of job assignments and key functions of various positions are provided as guidance in understanding a building department's structure and organization. Descriptions are also provided to demonstrate how to measure staff performance. The chapter also illustrates how to use staff performance data to fill staffing needs.

KEY POINTS:
- How is the makeup of the community related to the operation of the building department?
- What is a "combination inspector?"
- What is a "specialist inspector?"
- Why are organizational charts helpful?
- Why is it important to identify the responsibilities (key functions) of a building department?
- Although building departments may be structured differently throughout the country, they all generally have similar responsibilities. What are four major components of every building department?
- What is the difference between management and administrative functions?
- What responsibilities or functions are considered administrative?
- Besides reviewing plans, what other responsibilities are performed by the Plan Check division?
- What is a "performance measure?"
- Why should efficiency standards for quality and quantity be established?
- What quality standards apply to a building department?
- What is the most efficient way to track quantitative standards?
- What are four quantitative standards typical to every building department?
- What is the importance of determining adequate department staffing?
- What difficulties or "intangibles" are encountered when determining staffing needs of a department?
- How do public relations impact staffing decisions?

- Should future hiring decisions be based solely on the level of permit activity? Why or why not?
- What factors determine whether the department should hire combination inspectors or specialist inspectors?
- What is the value of performing "time and motion studies" on inspector activities?
- How does the "rule of thumb application" help in determining staffing levels?
- In determining the time available for inspections, what factors must be considered?
- How does the actual inspection workload influence staffing needs?
- What is the value of plan examination to developers and contractors?
- What nine unusual or adverse conditions warrant extra attention in the design of buildings?
- What qualifications should a plan examiner have?
- Why is it important to have qualified persons performing plan review?
- How does a building department determine the number of plan examiners needed?
- What staff positions are typically required in a building department?
- What is the relationship between permit revenues and a department's workload?
- How much of a building department's operating expenses should be paid through permit fees?
- How much of a building department's operating expenses should be paid from the jurisdiction's general fund?
- What is the value of examining multiple years permit revenue?
- What nonrevenue producing services do most building departments offer?
- How do nonrevenue producing services affect staffing needs?
- Why is it important to identify and track nonrevenue producing services?
- Are interpersonal skills essential for building department employees?
- What's the difference between inspecting by opinion and inspecting by the code?
- How does an inspector's demeanor influence or affect public relations?
- What are two essential qualities of all inspectors?
- Why are clear job assignments important?
- Why should an administrator understand personality traits of his/her employees?
- What factors should be included in a management precept?
- What is the importance of defining functions for every individual position of the department?

Quiz

Chapter 11

Multiple Choice

1. The structure of a building department should be based on _____.

    a. the demands of local contractors

    b. the makeup of the local community

    c. the projected revenue

    d. the industry standards

    Reference_____

2. A visual depiction of a building department's structure is called a(an) _____.

    a. linear diagram

    b. job responsibility chart

    c. key function

    d. organizational chart

    Reference_____

3. Which of the following key functions are typically performed by a building department?

    a. Administration/management

    b. Plan Review

    c. Inspections

    d. All of the above

    Reference_____

4. Automating records and other IT-related tasks are typically considered what type of function?

    a. Administrative

    b. Management

    c. Plan Review

    d. Inspection

    Reference_____

5. Performance measures are used to determine _____.

    a. quantity of permits issued

    b. type of permits issued

    c. quality and quantity of staff work activities

    d. Nonpersonnel related costs

    Reference_____

6. Which of the following would not be considered a "quantitative" performance measure?

    a. Public relations

    b. Phone service

    c. Plan review service

    d. Inspections

    Reference_____

7. Which of the following factors makes it difficult to precisely determine the staffing needs of a department?

    a. Permit counts

    b. Total revenues

    c. Maintaining good public relations and unforeseen, time-consuming problems

    d. Both a and b above

    Reference_____

8. To determine the average amount of time required for each inspection, _____.

    a. divide the total number of inspections per day by the number of inspectors available

    b. divide the total number of inspections per year by the number of inspector-hours

    c. multiply the nation-wide average by your regional modifier

    d. perform a time-and-motion study

    Reference_____

9. Which of the following is NOT a factor in establishing minimum number of inspectors needed for the department?

    a. Actual workloads

    b. Permit fees

    c. Travel time

    d. Type of building (ie., single-family, apartment complex, etc.)

    Reference_____

10. An example of overhead (nonrevenue producing) functions performed by a building department include _____.

    a) complaint investigation

    b) issuance and filing of permit records

    c) training personnel

    d) performing plan reviews

    Reference_____

11. An attribute that is frequently overlooked or undervalued when hiring staff for a building department is _____.

    a) trade experience

    b) education

    c) interpersonal skills personality traits

    d) job specific training and certification

    Reference_____

12. The decision of an inspector should always be based on _____.

    a) local construction practices

    b) code requirements and approved plans

    c) past experience, rules and procedures

    d) personal opinion

    Reference_____

13. To encourage employee growth and foster good public relations, the building official must _____.

    a) provide personnel with a clear understanding of their job functions

    b) recognize personality differences in personnel and manage these appropriately

    c) establish policies on how to administer the codes

    d) all of the above

    Reference_____

14. A list that establishes specific areas of responsibility and accountability is called a (an) _____.

    a) individual function list

    b) job description

    c) precept

    d) functional policy

    Reference_____

15. An obstacle to effective, consistent, comprehensive and fair code enforcement is/are _____.

    a) permit revenues

    b) poor public relations

    c) inadequate communication between the building department and local media

    d) lack of understanding of the reason for any given code requirement

    Reference_____

True/false

1. The building departments of all jurisdictions are structured the same.

    _____

    Reference_____

2. Inspectors should be regulators, not facilitators.

    _____

    Reference_____

3. It is not important for plan examiners to have licensure or certification in plan examination, architecture and/or engineering.

    _____

    Reference_____

4. Identifying key functions and individual functions will help an administrator define the duties and responsibilities of all employees.

    _____

    Reference_____

5. Individual responsibility lists should be static and nonchangeable.

    _____

    Reference_____

Completion

1. A 10-day turnaround time for plan review is an example of a

    _____.

    Reference_____

2. Measuring each step of every inspection and determining an average time to complete them is considered an example of _____.

    Reference_____

3. People who have training in a craft have a tendency to inspect by_____.

   Reference_____

4. Name three nonrevenue producing services typically performed by a building department.

   _____

   _____

   _____

   Reference_____

5. Name two of the four things good public relations depend on (the four f's).

   _____

   _____

   Reference_____

Chapter 12
The Public Counter

OBJECTIVES: To discuss and describe the role of the public counter in a building department.

LESSON NOTES: The first impression of a building department often happens at the public counter. The need for good public relations and consistent procedures is explained. Multiple suggestions are provided to enhance the public's understanding of the role and importance of the building department.

KEY POINTS:
- Why do public counter personnel need to understand the importance of exemplary customer service?
- Why do people resist getting a building permit?
- How does the public counter staff affect public relations?
- What type of education should be provided at the public counter?
- Why is it important to explain department procedures to the public?
- Who are typical building department customers?
- How and why does knowing the requirements for plan submittal impact customer service and public relations?
- What is the advantage in having trained public counter employees rather than plan examiners or building inspectors staff the counter?
- How can a building department effectively minimize confusion when the public counter becomes chaotic?
- What are typical steps involved in the permit application process?
- What is the value of performing a pre-permit inspection?
- What is the purpose of reviewing plans?
- How should the plan examiner deal with deficiencies in the plans?
- Once the permit is issued, how are plans distributed?
- What are the advantages and disadvantages of checking plans at the public counter?
- What important factors should be considered when checking plans at the public counter?
- How should permit fees be determined?
- What fundamentals must be understood in order to correctly assess fees based on valuations?

- Should the valuation of a structure exclude any components?
- How can disputes over valuation be resolved?
- What are the advantages of using a square-foot formula to determine fees?
- How do handouts benefit the building department?
- Why should the public counter staff understand the reasoning behind code provisions?
- What is the value of having established departmental procedures?
- When is it acceptable to deviate from established departmental procedures?
- What is the importance of "knowing your audience" when assisting customers at the public counter?
- How is construction terminology a help or hindrance to effective communication?
- How can a building department's website be used as a positive public relations tool?
- What information is helpful to provide on the department's website?
- What additional special services could be offered to enhance the building department's development and public relations?

Quiz

Chapter 12

Multiple Choice

1. Most often, the general public's first impression of their local building department
   _____.

   a. comes from television or newspaper

   b. is handed down by word-of-mouth

   c. occurs at the department's public counter

   d. is based on the department's website

   Reference _____

2. People visit the building department to obtain a permit because they _____.

   a. know a permit is required, and they want to comply with the law

   b. were told that they needed a permit, and they want to comply with the law

   c. got caught without one and are forced to comply with the law

   d. any of the above

   Reference _____

3. One of the main reasons the general public often resists getting permits is because
   _____.

   a. they think it will raise their property taxes

   b. they're lazy

   c. it's never mentioned on home-improvement television shows

   d. their parents never did

   Reference _____

4. The two most important elements of public counter service are public relations and
_____.

    a. permit fee collection

    b. public education

    c. plan examination

    d. interdepartmental referrals

    Reference _____

5. Public counter staff should _____.

    a. know exactly what information is needed to obtain a permit and be able to explain why

    b. be able to tell a joke in order to make the customer laugh

    c. tell a personal anecdote to put the customer at ease

    d. assist contractors before helping the general public

    Reference _____

6. The best way to inform people of submittal requirements and building department procedures for the most common construction projects is by _____.

    a. extensive, repetitive staff training

    b. inspection deficiency lists

    c. providing handouts and brochures

    d. press releases

    Reference _____

7. Checklists are helpful tools because _____.

    a. the customer can take them home for reference

    b. department staff can use them for consistency

    c. they provide details on department policy and/or procedures

    d. all of the above

    Reference _____

8. When a customer comes to the public counter, the staff should _____.

   a. acknowledge her/him promptly and courteously

   b. determine if he/she is an elected official

   c. check to see if someone else is available to help her/him

   d. wait until the customer asks for assistance

   Reference _____

9. To effectively handle a busy or chaotic public counter, the building department should _____.

   a. increase its budget to hire more staff

   b. fully train the public counter staff

   c. utilize handouts and checklists to maintain consistency

   d. both b and c

   Reference _____

10. If plan checking is provided at the public counter, the building department should restrict it to _____.

    a. engineered projects

    b. reviewing computations and calculations only

    c. projects not requiring action by other departments

    d. both a and b

    Reference _____

11. When the permit fee is based on valuation, the building department should _____.

    a. establish the valuation through an electoral ballot initiative

    b. base it on professional construction valuation methods

    c. require submittal of signed contracts for all permit applications

    d. establish the valuation by accepting local contractor estimates

    Reference _____

12. The building code states that the final building permit valuation must be set by
    _____.

    a. the building official

    b. local ordinance

    c. square-foot valuation tables

    d. none of the above; the code is silent on this issue

    Reference _____

13. To enhance public relations and promote a positive building department image, the
    public counter staff must _____.

    a. complete the application process in 5 minutes or less

    b. accept plans whether they are complete or not

    c. understand the importance of codes and be able to explain them to the customer

    d. cultivate a dispassionate demeanor

    Reference _____

14. Knowing when to use technical terminology and when to use lay terminology is
    _____.

    a. detailed in the administrative chapter of the building code

    b. an essential communication skill

    c. through quickly assessing the customer

    d. defined in the department operating procedures manual

    Reference _____

15. A building department can enhance their public relations by _____.

    a. having multilingual staff available

    b. providing an ADA-compliant work environment

    c. offering escrow accounts and/or accepting credit card payments

    d. any of the above

    Reference _____

True/false

1. Handouts and checklists are detrimental ways of providing information to the public.

    _____.

    Reference _____

2. Upholding the principles of public safety does not have to conflict with providing good customer service.

    _____

    Reference _____

3. It is important for public counter staff to know what is in the code and why.

    _____

    Reference _____

4. Building department personnel need to be able to assess their audience in order to speak to them at the appropriate level.

    _____

    Reference _____

5. A building department website leads to more phone inquiries because people don't understand what is on the site.

    _____

    Reference _____

Completion

1. A truly effective public relations and education program is intimately related to

    _____.

    Reference _____

2. Three steps involved in issuing virtually all permits are:

_____

_____

_____

Reference _____

3. Name two of the best ways to deal with a resentful, argumentative customer.

_____

_____

Reference _____

4. What are two things many citizens fear when permit fees are based on a valuation figure?

_____

_____

Reference _____

5. List three helpful information subjects to provide on the building department web site:

_____

_____

_____

Reference _____

Chapter 13
Using Information Technology in Building Departments

OBJECTIVES: To explain in more depth the value of computer and information technologies to building department functions.

LESSON NOTES: Many new technologies are outlined, including the pros and cons of each. The chapter provides clear steps to help the building department determine the best product for their needs.

KEY POINTS:
- How can technology help a building department?
- What departmental tasks can be managed with a customized permitting system?
- Why should new technology be implemented?
- What is an "enterprise" system?
- What are the three main types of electronic permitting systems?
- How does a Web component affect customer service?
- How does a Web component benefit the building department?
- Why should a permitting system match the needs and practices of the building department?
- What options are available with plan review software?
- How can new technology enhance a building department's inspection functions?
- What drawbacks or challenges do new technologies present to inspectors?
- What factors are involved in accepting permit fee payments on line?
- How can a permitting system help a building department develop reports and manage projects?
- What is the key factor in deciding whether a permitting system should be self-hosted or vendor-hosted?
- What are the three main options for providing system support and management?
- What functions should be detailed in a comprehensive service agreement?
- What factors influence a building department decision to implement new technology?
- What steps are involved in a successful implementation?
- How can a building department justify the cost of purchasing a new permitting system?

- What are the benefits of a department-level solution?
- What are the benefits of an enterprise solution?
- How should a request for proposals (RFP) be prepared?
- What steps should be taken when evaluating responses to an RFP?
- Why is it important to establish an implementation team?
- Who should be included on the implementation team?
- What are the main tasks to address when implementing a new system?
- What new permitting/inspection technologies are emerging?
- What are virtual inspections?

Quiz

Chapter 13

Multiple Choice

1. Building departments can streamline their permit process by _____.

    a. using a set of computer-based tools and services

    b. tracking plan review

    c. posting inspection results

    d. all of the above

    Reference _____

2. Which of the following is generally NOT a reason to implement new technology?

    a. To streamline existing processes

    b. To fix a specific problem

    c. To increase department revenue

    d. To improve the level of service to customers

    Reference _____

3. Electronic permitting was first implemented _____.

    a. by the U.S. military

    b. in 1975

    c. on mainframe computers with specialized software

    d. in jurisdictions with populations exceeding 500,000

    Reference _____

4. A permitting system comprising separate software systems for plan review, inspections, workflow and reporting is call a(n) _____ system.

    a. homegrown

    b. component

    c. integrated

    d. enterprise

    Reference _____

5. One module that allows customers to obtain information without speaking to department staff is a(n) _____.

    a. in-house network module

    b. mainframe network module

    c. workflow manager module

    d. interactive voice module

    Reference _____

6. Most local governments now use _____ to improve access and disseminate information.

    a. press releases

    b. web sites on the Internet

    c. local advertising agencies

    d. both b and c

    Reference _____

7. The main drawback for implementing electronic plan submittal and review is _____.

    a. cost

    b. terminology

    c. different codes adopted in different jurisdictions

    d. report functionality

    Reference _____

8. Backing-up department data is a vital service that provides _____.

    a. technical support

    b. system upgrades

    c. redundancy

    d. security

    Reference _____

9. The right solution for a department's permit needs should be evaluated and selected by _____.

    a. the elected officials

    b. the department head or building official

    c. the IT department

    d. a taskforce that includes building department and IT department personnel

    Reference _____

10. A(n) _____ may offer the best solution for a jurisdiction seeking to improve the services of multiple departments.

    a. prepackaged software system

    b. customizable software system

    c. IVR system

    d. enterprise system

    Reference _____

11. Proper management, clear communication, realistic expectations and knowledgeable decision making are keys to _____.

    a. a successful implementation of a new system

    b. choosing the right members of the implementation team

    c. securing adequate funding approval from elected officials

    d. managing a building department website

    Reference _____

12. Critical to the success of an implementation team is/are_____.

    a. adequate time allocated for team members to work with the vendor

    b. strict adherence to the contract terms

    c. the establishment of clear lines of authority amongst team members

    d. both a and c

Reference _____

13. Moving permit data from one system to another is _____.

    a. required with independent operating systems

    b. called "historical exchange"

    c. called "data migration" or "database migration"

    d. no longer an IT function/responsibility

Reference _____

14. A portable device that brings new technology to the inspectors is the _____.

    a) digital handwriting pen

    b) video stream

    c) LCD display

    d) all of the above

Reference _____

15. Implementing a system to perform virtual inspections can benefit a building department by _____.

    a) reducing overall travel time in the department

    b) allowing deployment of less highly trained staff in peak periods

    c) transmitting additional plan information between the office and the field (inspection) site

    d) all of the above

Reference _____

True/false

1. Implementing the right permitting software can lead to higher quality inspections.

   _____

   Reference _____

2. Many permitting systems can be enhanced by adding a voice or web component.

   _____

   Reference _____

3. Most building departments are moving towards accepting electronic plan submittals.

   _____

   Reference _____

4. The first step in implementing new technology is preparing a request for proposals.

   _____

   Reference _____

5. Data migration is a time-consuming but simple part of implementing a new system.

   _____

   Reference _____

Completion

1. When implementing new technology, the building official must understand its value to_____ and _____.

   Reference _____

2. Name two issues that are part of the intent of electronic permitting.

   _____

   _____

   Reference _____

3. Electronic permitting systems, when available on the web, provide value to the building department's customers by allowing them to_____ and _____.

    Reference _____

4. Name three elements of an effective RFP.

    _____

    _____

    _____

    Reference _____

5. Current and future use of remote video imaging is roughly tied to_____.

    Reference _____

Chapter 14
Records Management

OBJECTIVES: To provide an explanation of the need for and management of the various records used by a building department.

LESSON NOTES: The differences between public and nonpublic, permanent and temporary records is explained. Examples are provided to illustrate permit applications, plan review and inspector records, and departmental activity reports.

KEY POINTS:
- Why do governmental agencies keep records?
- What are transitory records?
- Are permanent records the same as public records?
- What records can be considered nonpublic?
- How and when is it permissible to dispose of records?
- What factors should be addressed to ensure accuracy of department records?
- Why do permit forms vary from jurisdiction to jurisdiction?
- What vital information should be included on all permit application forms?
- Of what importance or value is the design of permit application forms?
- Should permit applications include a signature line?
- How has on-line permitting impacted permit application forms?
- What are the advantages and disadvantages of issuing separate mechanical, electrical and plumbing permits for a single project?
- What types of reports are helpful in performing plan reviews?
- What are the disadvantages of relying on check lists when performing plan reviews?
- What are the advantages of utilizing form letters?
- What forms are used to record and perform inspections?
- What considerations should be given to the design and use of correction notices?
- Why should a building department be concerned with the design of their Certificate of Occupancy forms?
- How do annual reports benefit a building department?
- What information should be included in the building department's annual reports?
- How do diagrams enhance reporting?
- What public agencies utilize building department reports?
- How can building departments make their records available to the public?

Quiz

Chapter 14

Multiple Choice

1. One indicator of the effectiveness of a building department's administration is_____.

    a. the number of inspectors per trade

    b. the number of permits issued

    c. the efficiency of its record-keeping system

    d. the uniformity of its inspection and reporting records

    Reference _____

2. Which of the following is an example of a transitory record?

    a. Permit application forms

    b. Blueprints or construction plans

    c. Inspection request forms

    d. Private correspondences that do not reference a specific project

    Reference _____

3. Before disposing of public records, the building official should _____.

    a. verify proper procedures with the jurisdiction's legal counsel

    b. make copies for the local historical society

    c. publish copies in the local daily record

    d. none of the above; public records cannot be destroyed

    Reference _____

4. To safeguard the accuracy of computer records, only _____ should have the authority to amend or make changes to the records.

    a. the building official

    b. the clerk of the jurisdiction

    c. inspectors

    d. selected, highly trusted staff

    Reference _____

5. Local jurisdictions tend to develop forms based on _____.

    a. local needs

    b. past experiences

    c. past legal problems

    d. all of the above

    Reference _____

6. Which of the following is frequently NOT a required field of information on permit applications?

    a. Address of the proposed work

    b. Valuation

    c. Applicant contact information

    d. Description of work to be performed

    Reference _____

7. The main difficulty associated with issuing a single, combination building permit for a project is _____ .

    a. cost

    b. accounting for contractor licensing requirements

    c. inspection records

    d. both a and b

    Reference _____

8. Permit application forms _____.

    a) should be revised and updated on a periodic basis

    b) should never be revised

    c) should be prepared by the jurisdiction's law department

    d) should be formatted by the local legislative body

    Reference _____

9. Maintaining daily plan-check lists that document information such as number of plans checked _____.

    a. helps the building official prepare department reports

    b. can be used to predict inspection activities

    c. can guide the department's budget needs

    d. any or all of the above

    Reference _____

10. Many jurisdictions use _____ to expedite correspondence.

    a. administrators

    b. plan examiners

    c. form letters

    d. diagrams

    Reference _____

11. The final building department approval of a project is generally considered the _____.

    a. issuance of a business license

    b. permit closure

    c. issuance of a certificate of occupancy

    d. property release

    Reference _____

12. Legislative bodies gain an understanding of the functions of the building department through _____.

    a. annual department reports

    b. statistical analyses prepared by local media

    c. complaint procedures

    d. electronic permit software

    Reference _____

13. Rarely included but helpful information that a building department could add to its annual report are _____.

    a. explanations of any legal actions

    b. long-range goals

    c. short-term goals

    d. all of the above

    Reference _____

14. _____ tend to be the best way to visually illustrate trends over a period of time.

    a) Pie charts

    b) Line graphs

    c) Bar graphs

    d) Diagrams

    Reference _____

15. When a report includes problems that inhibit proper administration of the department, care should be exercised so that this information _____.

    a) subjugates the department

    b) dominates the report

    c) provides or offers solutions to the problems

    d) none of the above; reports should never include problems

    Reference _____

True/false

1. The only purpose of creating records is to document current activity.

   _____

   Reference _____

2. Copyrighted plans are not open to public view.

   _____

   Reference _____

3. Applicant signatures may not be required in jurisdictions that utilize web-based permitting.

   _____

   Reference _____

4. Using a single, combination permit reduces permit-processing time for the building department.

   _____

   Reference _____

5. The appearance and formatting of the Certificate of Occupancy is irrelevant.

   _____

   Reference _____

Completion

1. Name four different types of records.

   _____

   _____

   _____

   _____

   Reference _____

2. Two positive aspects of using check sheets for plan review are:

_____

_____

Reference _____

3. Five forms typically used by inspectors are:

_____

_____

_____

_____

_____

Reference _____

4. Three reasons to have attractive Certificate of Occupancy documents are:

_____

_____

_____

Reference _____

5. Four statistics typically included in the building department's annual report are:

_____

_____

_____

_____

Reference _____

15

Chapter 15
Customer Relations

OBJECTIVES: To describe the value to the building department of offering good public relations to all of its customers.

LESSON NOTES: Building department employees interact with a wide variety of individuals such as contractors, homeowners, elected officials and other jurisdictional agencies. This chapter provides examples of how to specifically address the needs of these various groups, resulting in positive public relations for the building department.

KEY POINTS:
- What is the difference between public relations and special relations?
- How do technical competence and public relations affect a building department's image?
- In what ways can an inspector impact the public's opinion of the building department?
- How should inspectors handle controversies on construction sites?
- Where is the line between a gift given in appreciation and one given as a bribe?
- Why is publicity about code enforcement necessary?
- What factors should be considered when determining the level of engineering services offered by the building department staff?
- Should building officials recommend professionals to the members of the public?
- Should building officials recommend specific products or materials?
- What are the most common criticisms of building departments by architects and engineers?
- How should the building official respond to complaints from the public about departmental procedures?
- How should the building official respond to complaints from architects or engineers about departmental procedures?
- What are the three major areas affecting relations between contractors and the building department?
- What are the various attitudes of building department staff towards contractors and builders?
- How friendly should building department staff be towards construction trade personnel?

- How should building departments treat gratuities?
- How can a building official determine product compliance?
- What is the ICC-ES, and how can it help building officials?
- What is a listing agency?
- What support do building craft organizations offer to building departments?
- Why is it important for the building official to maintain contact and good rapport with elected officials?
- What is the relationship between the building department and the fire department?
- How do the human relations and building departments assist each other in the hiring process?
- What is the importance of establishing a good relationship with the local media?
- What type of employee identification is appropriate for building department personnel?
- How should the building department effectively handle complaints?
- What is the value of building officials associating with their peers?

Quiz

Chapter 15

Multiple Choice

1. The wide variety of persons who contact the building department make it necessary for personnel to develop and understand the need for _____.

    a. individualized or special relations

    b. preferential considerations

    c. strict procedures

    d. a public relations division

    Reference _____

2. Good public relations between department personnel and the public can be achieved _____.

    a. through correspondence courses

    b. by careful and deliberate planning

    c. by training staff in developing human relations skills

    d. either or both b and/or c

    Reference _____

3. Inspectors can handle jobsite controversies by _____.

   a. explaining the reason and the basis of the code requirement to the foreman

   b. understanding that the issue is not personal and finding a good way to explain the requirement to the foreman

   c. asking to speak to the foreman's superior if the foreman is not convinced and is not in agreement with the requirement

   d. all of the above

   Reference _____

4. The general public usually does not know that an inspector _____.

   a. prefers to receive gifts for performing the inspection

   b. is required to meet the contractor on every inspection

   c. is allowed to meet a homeowner by appointment only

   d. is a law enforcement agent

   Reference _____

5. Observing an annual "Building Safety Week" _____.

   a. promotes public awareness of the value and importance of codes

   b. provides the building department an opportunity to educate the public on the role and function of the department

   c. is an excellent time for speaking to community groups

   d. all of the above

   Reference _____

6. The building department will be viewed as playing favoritism when _____.

   a. recommending an architect to a homeowner

   b. providing a short list of architects to a homeowner

   c. disallowing a product that is new to the market for lack of an evaluation report

   d. both a and b

   Reference _____

7. Building officials should rely on _____ for technical expertise during code adoptions.

   a) elected officials

   b) local business and community leaders

   c) local architects and engineers

   d) building department personnel only

   Reference _____

8. A poll of architects and engineers showed praise for building departments that _____.

   a. did not allow preliminary plan review conferences

   b. did not assign priority to plan review based on project complexity

   c. provided smooth processing of plans in a reasonable time period

   d. both a and b

   Reference _____

9. A completely benevolent attitude of building department personnel toward the trades is _____.

   a. unacceptable

   b. completely innocuous

   c. proof of collusion

   d. the goal of all public relations training programs

   Reference _____

10. Which of the following is relied upon by the inspector for product evaluation or listing?

    a. ICC-Evaluation Service (ICC-ES)

    b. Underwriters Laboratories (UL)

    c. American Gas Association (AGA)

    d. all of the above

    Reference _____

11. When dealing with elected officials, a wise building official should _____.

    a. explain the impact of changes to long standing practices and requirements

    b. provide them with an understanding of the building department's purpose and function

    c. always display a friendly, competent, respectful demeanor

    d. all of the above

    Reference _____

12. The phrase "You build them and we live with them" represents the relationship between the building department and the _____.

    a. local housing authority

    b. human relations department

    c. fire department

    d. elected officials

    Reference _____

13. To enhance public relations, one of the most important on-going responsibilities of every building official is _____.

    a. the adoption of new codes

    b. the hiring process

    c. promoting the importance of the department through public education

    d. both a and c

    Reference _____

14. A building department's public image will be MOST enhanced by _____.

    a. publication of technical manuals

    b. responding to complaints in a timely, organized manner

    c. establishment of and adherence to strict procedures

    d. providing uniforms to all personnel

    Reference _____

15. In a dispute between an owner and a contractor, the inspector is safe _____.

    a. to always support the owner

    b. to always support the contractor

    c. to not get involved in such disputes to begin with

    d. to review the disputed construction for code compliance

    Reference _____

True/false

1. To be a successful building official, one must have specialized training in personality traits.

    _____

    Reference _____

2. The expertise of architects and engineers can provide valuable assistance to a building department.

    _____

    Reference _____

3. The general public cannot be expected to understand all the complexities of construction projects.

    _____

    Reference _____

4. Responding to complaints should not be a priority for any building department.

    _____

    Reference _____

5. Because each building department is unique, there is no value in networking with other building departments.

    _____

    Reference _____

Completion

1. Name the four steps in effectively handling complaints.

    _____

    _____

    _____

    _____

    Reference _____

2. The best attitude for a building department employee to adopt when dealing with the public is _____.

    Reference _____

3. Name three ways for a building department to educate the public in the need for and validity of codes.

    _____

    _____

    _____

    Reference _____

4. Having a clearly defined complaint procedure may not reduce the number of complaint calls, but it will _____.

    Reference _____

5. The greatest compliment that construction professionals can give their local building department is _____.

    Reference _____

Chapter 16

Legal Aspects of Code Administration

OBJECTIVES: To become familiar with basic principles of law and legal terms applicable to administration of a building department.

LESSON NOTES: This chapter introduces the federal, state and local legal issues that must be considered by code administrators. It discusses duties of the building official, sovereign immunity, legal due process, legal liability and several other topics and related subjects.

KEY POINTS:
- What are discretionary and ministerial acts?
- How are deviations from code requirements processed or allowed?
- What is the purpose of the board of appeals?
- The basic sources of law are from which documents?
- Dillon's rule deals with which principle?
- What does the doctrine of preemption state?
- How are common law and statutory laws different?
- What does the doctrine of sovereign immunity apply to?
- What is the role of the federal government in sovereign immunity?
- What are special districts, and what is their main purpose?
- What are some examples of public nuisance?
- How is a building department vulnerable to legal attack?
- How is procedural due process intended to protect the rights of parties against government action?
- How does substantive due process protect constitutional rights?
- What issues does inverse condemnation deal with, and how does it relate to building department functions?
- How can legal problems be prevented or reduced?
- At which stage of the legal proceedings does plea bargaining become a useful tool?
- Based on which types of penalty are most building codes enforced?
- Which part of the U.S. Constitution governs inspections?
- Under what circumstances could warrantless inspections be performed?
- To what extent are governments and building officials held liable for negligence?

- What are the elements that a plaintiff must prove to prevail on a claim of negligence?
- How does a plaintiff prove a civil rights claim?
- What are some examples of shortcomings in local adoption of model codes?
- What are the elements that a model building regulation act should include?
- What is the purpose of the statute of limitation?

Quiz

Chapter 16

Multiple Choice

1. Building officials are _____.

   a. construction coordinators

   b. legal advisors

   c. law enforcement officers

   d. none of the above

   Reference _____

2. Two major components of the legal aspects of building department administration are _____.

   a. customer service and timely issuance of permits

   b. plan review and inspections

   c. employee training and employee evaluation

   d. enforcement and prosecution

   Reference _____

3. Punishments for code violations are determined by _____.

   a. the building official

   b. the courts

   c. the board of appeals

   d. jurisdiction's attorney

   Reference _____

4. Which of the following has the authority to grant a code modification if the requirement seems to be ludicrous?

    a. Building official

    b. Board of appeals

    c. Inspector

    d. None of the above

    Reference _____

5. The legal principal that prevents local building departments from regulating buildings owned by the federal government is _____.

    a. sovereign immunity

    b. preemption

    c. due process

    d. all of the above

    Reference _____

6. Of great help to the jurisdiction legal council is _____.

    a. when the building official does not get involved in legal discussions at all

    b. when the building official takes the leading role in jurisdiction's legal issues

    c. when the building official is very knowledgeable on legal issues specifically related to building department matters

    d. none of the above

    Reference _____

7. Taking or damaging private property without just compensation is _____.

    a. inverse condemnation

    b. forceful entry

    c. preemption

    d. both b and c

    Reference _____

8. Covenants and easements that are not enforced by local building departments are: _____.

    a. engineering department public agreements

    b. state required public agreements

    c. private agreements

    d. agreements between the building department and private property owners

    Reference _____

9. Approval of building permits can be delayed by building departments until _____.

    a. property owner corrects unsafe conditions on another site

    b. property owner certifies that construction will begin 180 days after the permit is issued

    c. the applicant corrects the plans in accordance with all code deviations listed by the plan check staff

    d. all of the above

    Reference _____

10. An acceptable manner of determining a property owner to whom a notice of violation should be addressed is _____.

    a. the local telephone book

    b. tax records

    c. by calling the neighbors

    d. the most recent deed

    Reference _____

11. If the building official request to enter the premises for inspection is denied, the building official may enter for inspection only after _____.

    a. the occupant is reminded that the building official is a law enforcement officer

    b. an administrative search warrant is issued

    c. the building official threatens that he/she will seek a search warrant

    d. none of the above

    Reference _____

12. Which of the following conditions is not an acceptable justification for a warrantless search?

    a. Emergency or exigent situation

    b. Observation of a violation from a public place

    c. City-wide rental housing inspection program authorized by the City Council and Mayor

    d. Consent of apartment tenant

    Reference _____

13. In a claim of negligence against a government official, what is the standard of proof the plaintiff must meet? They must prove all element of their case _____.

    a. by a preponderance of evidence

    b. beyond a reasonable doubt

    c. both a and b

    d. neither a nor b

    Reference _____

14. If a contractor begins work on a project that requires a permit without obtaining the permit, when does the period of the statute of limitations begin?

    a. When the work begins

    b. When the project is completed

    c. 180 days after the date permit was issued

    d. None of the above

    Reference _____

15. Camara vs. Municipal Court of the City and County of San Francisco was a case regarding _____.

    a) proper collection of evidence

    b) exigent circumstances

    c) right of entry

    d) covenants and easements

    Reference _____

True/false

1. The discretionary authority given to building officials is usually quite limited.

    _____

    Reference _____

2. Under the provisions of the *International Building Code* (IBC), the building official is never authorized to grant modifications, in individual cases, to the requirements of the code.

    _____

    Reference _____

3. Under the doctrine of preemption, a municipality of county may not pass a law that is inconsistent with state law.

    _____

    Reference _____

4. "Special Districts" formed for the purpose of providing sanitary facilities are quasi-government agencies.

    _____

    Reference _____

5. Prima Facie means a friend of the court.

    _____

    Reference _____

Completion

1. An _____ is a privilege, or right, that one person has in the land of another without paying for the exercise of it.

    Reference _____

2. The authority of control is derived from the _____ granted to governmental entities by constitutional provisions and is basic to the premise of the consent of the people to be governed.

    Reference _____

3. Because building officials work for state, county or municipal agencies, any action taken by them is a _____.

   Reference _____

4. The consent of the occupant for entry must be given _____.

   Reference _____

5. A statute of limitation is a _____ established by the legislative body within which actions against persons violating laws must be initiated.

   Reference _____

Chapter 17
Disaster Mitigation and Building Security

OBJECTIVES: To describe the many mitigation programs and tools available to help jurisdictions reduce the risks associated with natural and man-made disasters.

LESSON NOTES: The benefits of disaster mitigation strategies and preparedness are thoroughly discussed. Numerous federally funded programs are described. The effect of man-made disasters on emergency preparedness is also detailed.

KEY POINTS:
- What has been the impact on human life and property from recent natural disasters?
- How can governments better protect lives and property within their communities?
- What is disaster mitigation?
- What is the economic benefit of implementing disaster mitigation strategies?
- What is the Stafford Act?
- How did the Volkmer Amendment to the Stafford Act impact funding?
- What is the national Flood Insurance Reform Act of 1994?
- What changes in federal funding for disaster mitigation have taken place since 2001?
- What are the four major categories of mitigation activities?
- How do modern building codes help mitigate risk?
- How does land-use planning help mitigate risk and reduce loss or damage?
- How can local governments use their capital improvement plan as a mitigation tool?
- What is the value of redundancy in a capital improvement plan?
- What are the pros and cons of hazard control methods such as levees and dams?
- What funding is available to support state and local disaster mitigation efforts?
- What are Hazard Mitigation Grant Program funds used for?
- What types of grants are available through the Flood Mitigation Assistance Program?
- How does the Pre-Disaster Mitigation Grant Program assist hazard mitigation activities?
- Why did FEMA develop the initiative to make universities more disaster resistant?
- What federal technical assistance programs are available to assist local mitigation efforts?
- What programs are available to help reduce and mitigate high-wind hazards?

- How can wind hazards be mitigated through nonstructural measures?
- What programs are available to reduce and mitigate seismic hazards?
- What structural and nonstructural changes can be made to retrofit buildings in order to mitigate seismic hazards?
- What are the goals of the National Earthquake Hazards Reduction Program?
- How can houses be modified to mitigate flood and mudslide hazards?
- What constitutes a wildland-urban interface?
- What measures can be taken to mitigate wildland-urban fires?
- How can families prepare for winter storms?
- How have terrorist attacks impacted disaster mitigation planning and programs?

Quiz

Chapter 17

Multiple Choice

1. The World Watch Institutes report in 2001 states that _____ alone cause nearly ⅓ of all economic losses and ½ of all deaths.

    a) hurricanes

    b) tornadoes

    c) floods

    d) earthquakes

    Reference _____

2. Sustained action taken to reduce long-term risks to human life and property from a hazard event is called _____.

    a) damage assessment

    b) disaster mitigation

    c) disaster recovery

    d) organizational planning

    Reference _____

3. Restricting floodplain areas for low-risk uses such as parks or golf courses is an example of _____ .

    a) land-use planning

    b) organizational planning

    c) hazard control

    d) design construction

    Reference _____

4. Many communities now view_____ as a last-resort method of hazard control owing to their potential for failure and continued maintenance expense.

    a) levees

    b) dams

    c) retaining walls

    d) all of the above

    Reference _____

5. The federal government supports state and local mitigation efforts by
    _____ .

    a) offering grant funding

    b) offering technical assistance

    c) limiting Pre-Disaster Mitigation Funding

    d) both a and b

    Reference _____

6. Safe rooms and in-home shelters are designed to offer protection from which natural hazard?

    a) tornadoes and hurricanes (high winds)

    b) earthquakes

    c) floods

    d) ice and snow

    Reference _____

7. Which of the following is considered a structural retrofit for mitigating damage from earthquakes?

    a) Between-joist blocking and bridging

    b) Lateral bracing of cripple walls

    c) Anchoring the wood frame to the foundation

    d) All of the above

    Reference _____

8. Defensive measures used to reduce the vulnerability of people and property to terrorist acts is called _____ .

    a. antiterrorism

    b. counterterrorism

    c. mitigation planning

    d. family preparedness planning

    Reference _____

9. A design tool that can be used to mitigate both natural and man-made disasters is/are _____ .

    a. backflow valves

    b. spark arrestors on chimneys

    c. structural hardening

    d. both b and c

    Reference _____

10. FEMA's Hazard Mitigation Grant Program was established by _____ .

    a. constitutional amendment

    b. the Stafford Act, 1988

    c. the Volkmer Amendment, 1993

    d. Dillon's Rule

    Reference _____

11. The Tornado Shelters Act specifically authorized the use of Community Development Block Grant (CDBG) funds to construct tornado-safe shelters in _____ .

    a. colleges and universities

    b. manufactured-home parks

    c. schools

    d. all of the above

    Reference _____

12. The process of mitigating terrorist hazards before they become disasters involves the need to _____.

    a. identify and organize resources

    b. conduct a risk or threat assessment

    c. identify mitigation measures

    d. all of the above

    Reference _____

13. In preparation for ice, snow and winter storms, families should typically have emergency supplies, food and medications for at least _____.

    a. one week

    b. one day

    c. one month

    d. two months

    Reference _____

14. Federal agencies that are part of the National Earthquake Hazards Reduction Program (NEHRP) are _____.

    a. the Federal Emergency Management Agency (FEMA)

    b. the National Institute of Standards and Technology (NIST)

    c. the National Science Foundation (NSF)

    d. all of the above

    Reference _____

15. Water heaters are generally protected from sliding and falling over in a seismic event by _____.

    a. being strapped to walls

    b. being placed in a closet

    c. by being placed tight against storage racks and furniture

    d. all of the above

    Reference _____

True/false

1. Natural disasters can have long-term economic and social impacts on the communities in which the disaster occurred.

    _____

    Reference _____

2. Disaster mitigation has proven to be cost effective.

    _____

    Reference _____

3. Adopting and enforcing modern building codes is the most cost-effective way to mitigate risks.

    _____

    Reference _____

4. Funds from the Hazard Mitigation Grant Program may be used to cover the costs of elevating homes in flood-prone areas.

    _____

    Reference _____

5. All earthquake mitigation measures are structural.

    _____

    Reference _____

Completion

1. Name four major categories of mitigation activities.

    _____

    _____

    _____

    _____

    Reference _____

2. Land-use planning is most effective in _____ or _____.

    Reference _____

3. Name at least three federal grant programs that support state and local mitigation efforts.

    _____

    _____

    _____

    Reference _____

4. The National Hurricane Program provides three categories of assistance and support. List them.

    _____

    _____

    _____

    Reference _____

5. Homes located in or near wildland areas can mitigate damage from wildfires by _____ and _____.

    Reference _____

Chapter 18
Housing, Property Maintenance and Code Enforcement Inspection Programs

OBJECTIVES: To provide methods, options and the many facets of housing and property maintenance inspection programs and assess their impact on the community.

LESSON NOTES: Federal legislation and model housing codes are tools to help local communities address the need to provide safe living environments for their residents. Essential components of an effective housing inspection program are delineated. Examples are provided highlighting the results in selected communities.

KEY POINTS:
- How did population growth lead to public health concerns in the United States?
- How did the U.S. Housing Act affect substandard housing in the late 1930s and again in 1949?
- What is the objective of a housing code?
- Why should a housing code or property maintenance code exceed a minimal level of standards?
- How has the document "A Proposed Housing Ordinance," published in 1952, influenced the development of housing codes in the United States?
- What is the difference between *housing* codes and *building* codes?
- What essential elements should be included in any housing code?
- Why is decay and deterioration of tenements and apartment houses more problematic than decay and deterioration of individual dwellings?
- What is *urban blight*?
- What is the economic impact of blighted areas on a community?
- How can jurisdictions address the challenge of controlling blight?
- What are the pros and cons of *reactive* property maintenance inspection programs?
- What elements are associated with successful *proactive* property maintenance inspection programs?
- How should a community assess the need for establishing a property maintenance inspection program?
- How does a community's legislative body affect the implementation of a property maintenance inspection program?

- Why are initiatives and performance measures important to a successful enforcement program?
- What preventative measures can a jurisdiction take to enhance the effectiveness of its code enforcement program?
- How has the recent shortage of low-rent housing affected code enforcement programs?
- What legal authority do jurisdictions have to abate substandard buildings?
- What precautions should be taken, and what procedures should be followed when a jurisdiction takes *summary action* against a substandard building?
- How can a multiple-family dwelling inspection program benefit a community?
- How can a proactive single-family inspection program benefit a community?
- How can a single-family rental inspection program benefit a community?
- What role can neighborhood groups play in reducing substandard housing?
- What is the importance of due process in resolving substandard conditions?
- What processes can a jurisdiction employ to resolve noncompliant cases?
- What are the attributes of an effective code enforcement program?

Quiz

Chapter 18

Multiple Choice

1. In addition to police and fire protection, local governments foster safe living conditions for their residents through _____.

    a. comprehensive community planning

    b. adoption of building and construction codes

    c. adoption of property maintenance and housing standards

    d. all of the above

    Reference _____

2. Property maintenance codes are, essentially, _____.

    a. unenforceable guidelines

    b. an environmental health protection code

    c. an urban aesthetic code

    d. a panacea

    Reference _____

3. Until 1967, the fundamental reference for housing legislation was _____.

    a. the 1867 Tenement Housing Act

    b. the Housing Act of 1949

    c. the document "A Proposed Housing Ordinance"

    d. the U.S. Department of Housing and Urban Development

    Reference _____

4. Which of the following would be considered a *substantive provision* of a housing code?

   a. Adequate maintenance of the building

   b. Adequate maintenance of the grounds and property

   c. Minimum facilities and equipment in a dwelling unit

   d. All of the above

   Reference _____

5. Dilapidated buildings, graffiti, litter and broken sidewalks are some of the signs of _____.

   a. poverty

   b. urban blight

   c. homelessness

   d. economic turmoil

   Reference _____

6. To be successful, local, proactive property maintenance programs must be supported by _____.

   a. the legislative body

   b. local tax revenues

   c. state and federal tax revenues

   d. grant funding

   Reference _____

7. The _____ requires cities to have comprehensive area inspection programs in order to remain eligible for national renewal funds.

   a. Proposed Housing Ordinance of 1952

   b. *International Property Maintenance Code* (IPMC)

   c. Housing Act of 1949

   d. Housing Act of 1964, Section 301

   Reference _____

8. Whether or not a community should establish a property maintenance program is typically determined _____.

    a. by assessment of existing housing and property conditions

    b. through ballot initiatives

    c. by the building official

    d. by the police and/or fire departments

    Reference _____

9. The success of a jurisdiction's code enforcement program can be best illustrated by _____.

    a. local media coverage

    b. legislative resolution

    c. specific goals

    d. performance measures

    Reference _____

10. Which of the following is not one of the general categories for achieving compliance in a property maintenance program?

    a. Resolution

    b. Performance measures

    c. Intervention

    d. Prevention

    Reference _____

11. When large areas of dilapidated buildings need to be demolished, this effort is handled more effectively by _____.

    a. housing authorities

    b. renewal agencies

    c. building departments

    d. either a or b

    Reference _____

12. _____ is a term used to describe the decision to order a building to be immediately demolished because it poses an imminent danger to the public.

   a. Due process

   b. Eminent domain

   c. Summary action

   d. Intervention

   Reference _____

13. Due process does NOT require _____.

   a. hand-delivered notification

   b. legal notification of the proper party

   c. clarity with respect to the alleged violation

   d. specific details of the alleged violation

   Reference _____

14. The document that serves as a legal notice to the owner or tenant to correct code deficiencies is typically referred to as a _____.

   a. daily report form

   b. court action form

   c. violation notice

   d. no-entry notice

   Reference _____

15. An effective code enforcement program _____.

   a. does not require an attorney for court appearances

   b. does not address weeds, litter or junk cars

   c. has a strategic plan with definite, measurable goals

   d. establishes new construction standards in the community

   Reference _____

True/false

1. If urban blight is to be controlled, housing codes must be administered throughout an entire community.

   _____

   Reference _____

2. If a housing code causes undue hardship on some individuals, it should be abandoned and not enforced.

   _____

   Reference _____

3. Blight is self-generating.

   _____

   Reference _____

4. Most communities initiate a property maintenance inspection program as a reactive response to complaints.

   _____

   Reference _____

5. Jurisdictions typically receive more complaints about the condition of owner-occupied properties than about nonowner-occupied properties.

   _____

   Reference _____

Completion

1. With the enactment of _____, Congress declared that a decent home and suitable living environment is key to the general welfare and security of the nation.

   Reference _____

2. The objective of a housing code is _____.

   Reference _____

3. The primary difference between building and housing codes is that building codes define _____ and housing codes define _____.

    Reference _____

4. The four elements common to a successful, proactive property maintenance program are:

    _____

    _____

    _____

    _____

    Reference _____

5. When access for inspections is denied, the inspector should _____.

    Reference _____

19

Chapter 19
Building Sustainability: Preserving the Existing Residential Stock

OBJECTIVE: To describe issues associated with preserving and rehabilitating existing residential buildings and understand some of the historical obstacles in the way of effective rehabilitation of existing buildings.

LESSON NOTES: Economic, development, construction and occupancy barriers to building rehabilitation are defined and examined. Several programs and solutions are presented specifically addressing the challenges of this work.

KEY POINTS:
- How is rehabilitation defined, and what does it encompass?
- What is the impact of housing rehabilitation (rehab) on the U.S. economy?
- Are there obstacles unique to rehab as opposed to new construction?
- What are the general characteristics of the American housing profile?
- What are the three general categories of barriers related to various stages of renovation?
- What strategies reduce development stage barriers to rehab?
- What strategies reduce construction stage barriers to rehab?
- What strategies reduce occupancy stage barriers to rehab?
- Nationwide, what is the need for rehab?
- How is affordability measured?
- How does affordability impact housing rehab?
- How do development, construction and occupancy barriers interrelate and reinforce each other?
- How do best practice solutions address multiple barriers?
- What is the historical background of building code provisions for rehab?
- What challenges to rehab are created by the "25 – 50 percent rule"?
- What challenges to rehab are created by the "change of occupancy rule"?
- How did HUD's "Rehabilitation Guidelines" influence and change national building codes?
- What are some building code best practices for rehab?
- How and why has historic preservation evolved on the national level?

- How and why has historic preservation evolved at the local level?
- How does historical preservation contribute to housing rehab?
- In what ways is historical preservation a barrier to housing rehab?
- What are some historical preservation best practices for rehab?

Quiz

Chapter 19

Multiple Choice

1. In the U.S., approximately_____is spent on housing rehab annually.

    a. $10 – 20 million

    b. $100 – 200 million

    c. $10 – 20 billion

    d. $100-200 billion

    Reference _____

2. _____is/are defined as "a series of highly effective actions that help realize a specific objective."

    a. Housing rehabilitation

    b. Rehabilitation guidelines

    c. Best practices

    d. Design incentives

    Reference _____

3. Approximately _____ of housing units in the U.S. are single-family dwellings.

    a. 25 percent

    b. 52 percent

    c. 67 percent

    d. 71 percent

    Reference _____

4. Two occupancy stage barriers to rehab are _____.

    a. property tax and land use restrictions

    b. rent control and property taxes

    c. rent control and accessibility

    d. historic preservation and accessibility

    Reference _____

5. Housing affordability is estimated by figuring the _____.

    a. housing expense to income ratio

    b. remodel expense to energy savings ratio

    c. annual income to debt ratio

    d. all of the above

    Reference _____

6. Building codes have historically created problems for renovation on account of their emphasis on and orientation to _____.

    a. new construction

    b. accessibility

    c. fire-resistive construction

    d. occupancy classification

    Reference _____

7. The HUD _____ is the model document for developing rehab codes.

    a. Nationally Accredited Remodel and Renovation Program

    b. Nationally Applicable Recommended Rehabilitation Provision

    c. Nationally Applicable Recommended Resources Program

    d. Nationally Accredited Renovation Resource Package

    Reference _____

8. Many jurisdictions, including the following states have adopted a rehab code: _____.

    a. New York

    b. New Jersey

    c. Rhode Island

    d. all of the above

    Reference _____

9. _____ establishes a sliding scale of requirements depending on the level and scope of the rehab activity.

    a. Hazard classification

    b. Predictability

    c. Proportionality

    d. Affordability

    Reference _____

10. Preserving historic buildings was not a prevailing sentiment of American society until _____.

    a. the 1860s

    b. the 1920s

    c. the 1960s

    d. the 1970s

    Reference _____

11. Portions of which of the following federal acts support historic preservation?

    a. National Historic Preservation Act

    b. 1996 Transportation Act

    c. 1969 National Environmental Policy Act

    d. All of the above

    Reference _____

12. _____ may regulate actions by private owners of historic buildings.

    a. Local preservation commissions

    b. Federal regulations

    c. State regulations

    d. Either b or c

    Reference _____

13. Surveys estimate that _____ of all rehab projects are on properties designated as historic.

    a. 1 – 2 percent

    b. 5 – 10 percent

    c. 12 – 15 percent

    d. 17 percent

    Reference _____

14. The most prominent incentive to encourage historic renovation is the _____.

    a. historic tax credit

    b. federal income tax initiative

    c. state preservation loan program

    d. community tax credit

    Reference _____

15. Commonly used best practices for rehab of historically designated buildings include: _____.

    a) flexible treatment in the building codes

    b) enhanced application of historic tax credits

    c) early contact with all agencies regulating historic rehab

    d) all of the above

    Reference _____

True/false

1. Rehab activity constitutes about 2 percent of U. S. economic activity.

   _____

   Reference _____

2. Approximately ⅔ of U.S. households own rather than rent.

   _____

   Reference _____

3. Rehab activity is more predictable and easier to manage than new construction.

   _____

   Reference _____

4. Research has demonstrated that rehab/smart codes increase the cost of renovations.

   _____

   Reference _____

5. Federal interstate highway programs led to demolition of many historic areas.

   _____

   Reference _____

Completion

1. Approximately _____ of homes built between 1980 and 1995 require some form of rehab.

   Reference _____

2. Barriers to housing rehab are interrelated and_____.

   Reference _____

3. Retrofitting an existing building to the standards for new construction is often_____ and _____.

   Reference _____

4. _____ created a review process to evaluate federal undertakings that threaten National Register resources.

   Reference _____

5. The _____ of historic preservation often encourage and support rehab.

   Reference _____

20

Chapter 20
Rehabilitation and General Building Code Approaches

OBJECTIVES: To present various approaches to regulating repairs, alterations, change of occupancy and rehabilitation of existing buildings through modern rehabilitation codes such as the *International Existing Building Code* (IEBC).

LESSON NOTES: As communities strive to return existing buildings to a useable, occupiable condition, building officials have struggled to apply the provisions for new construction to these projects. Descriptions are provided for the new codes and standards that have been developed to address the unique conditions encountered during rehabilitation of existing buildings.

KEY POINTS:
- What are the differences between the terms alteration, repair and rehab?
- How does the term change of occupancy apply to existing buildings?
- What is the "25 – 50 percent rule"?
- Where are provisions for existing buildings found in the *International Building Code* (IBC)?
- How did the legacy code organizations (BOCA, ICBO and SBCCI) address the need for regulations specific to rehabilitation of existing buildings?
- Where can a building official find help and guidelines for issues related to the rehabilitation of existing buildings?
- What was the basis for the ICC's development of the *International Existing Buildings Code* (IEBC)?
- What is the intent of the IEBC?
- Why is it necessary to analyze an existing building before starting a rehabilitation project?
- After an analysis is completed, what are the next steps for review and approval of a rehabilitation project?
- How does the building official address the presence of archaic systems or materials in existing buildings?
- What is the impact of historic designations on rehabilitation projects?
- What are the unique administrative issues that a local jurisdiction should consider when managing rehabilitation proposals?

Quiz

Chapter 20

Multiple Choice

1. More existing buildings are being rehabilitated mainly as a result of increased interest in and support of _____.

   a. historical urbanism

   b. urban renewal

   c. federal grant funding

   d. disaster planning

   Reference _____

2. Repairs in general allow existing building materials to be replaced with the same materials, except for _____.

   a. lath and plaster

   b. slate shingles

   c. leaded stained glass windows

   d. glazing in hazardous locations

   Reference _____

3. The first step in a rehabilitation project is _____.

   a) analyzing the existing building's condition

   b) determining required safety features

   c) establishing a budget

   d) selecting a qualified contractor

   Reference _____

4. Retention of visual features may be mandatory when rehabilitating _____.

   a) high-rise buildings

   b) historical buildings

   c) civic buildings

   d) educational buildings

   Reference _____

5. In addition to national code organizations, which U.S. government entity has published reference documents for rehabilitation of existing buildings?

   a. Senate

   b. Supreme Court

   c. Department of Housing and Urban Development (HUD)

   d) Department of Health and Safety

   Reference _____

6. The basis for the "25 – 50 percent rule" was _____.

   a) early fire codes

   b) the insurance industry standards

   c) the prevalence of slums

   d) all of the above

   Reference _____

True/false

1. The term repair applies only to mechanical building systems. _____

   Reference _____

2. The term change of occupancy refers only to a change from one occupancy classification to another. _____

   Reference _____

3. The "25 – 50 percent rule" pertains to the percentage of square footage being altered in an existing building. _____

   Reference _____

4. The "Nationally Applicable Recommended Rehabilitation Provisions" (NARRP) document was the basis for the development of the IEBC. _____

   Reference _____

5. For major rehab projects, it is helpful to the building department to perform a field inspection prior to issuing the permit. _____

   Reference _____

6. Understanding how code requirements impact historic features will help the building official and design professional identify alternative compliance methods. _____

   Reference _____

Chapter 21
Green Building and Sustainability

OBJECTIVES: To present clear information and description of the impact of construction activity and buildings on regional and global environmental problems and the various programs and rating systems related to sustainable buildings.

LESSON NOTES: The concepts of green building and sustainability are discussed. Tables are provided to identify the benefits of sustainable building and of selected building components and systems.

KEY POINTS:
- How are building codes and standards related to sustainable building?
- What is the environmental impact of constructing and operating buildings?
- How do green buildings impact the environment?
- What is sustainable development?
- Why is it important to understand the relationship between humans and the natural environment?
- What are the fundamental principles of green building practices?
- What are the societal, economic and environmental benefits of sustainable building?
- What are the environmental consequences associated with the production and use of building materials?
- What is energy intensity as it pertains to building materials?
- What is life-cycle assessment?
- How do building materials impact indoor air quality?
- What are the environmental benefits of selected building components and systems?
- How will the current push for sustainable buildings impact the scope of building codes and standards?
- Who is promoting green building in the United States?
- What is the LEED standard?
- What standards currently exist related to building sustainability?
- What is "ISO 14040"?
- What agencies and organizations are working to develop green standards for inclusion into the ICC codes?

Quiz

Chapter 21

Multiple Choice

1. The intent of green or sustainable building is to _____.

   a. reduce the environmental impact of buildings

   b. enhance the use of new building materials

   c. improve the safety of the built environment

   d. all of the above

   Reference _____

2. To include provisions for sustainability, building codes must adopt a broader context of _____.

   a) product standards

   b) public welfare

   c) enforceability

   d) system choices

   Reference _____

3. More than _____ percent of the U.S. energy supply is used for the construction and operation of buildings.

   a. 20 percent

   b. 30 percent

   c. 40 percent

   d. 60 percent

   Reference _____

4. Development that meets the needs of the present without compromising the ability of future generations to meet their own needs is called _____.

    a. compassionate development

    b. judicious development

    c. sustainable development

    d. reliable development

    Reference _____

5. Which of the following is not a passive building strategy?

    a. Site orientation

    b. Window placement and shading

    c. Material performance

    d. Ground-source heat pumps

    Reference _____

6. The energy required to extract, process and manufacture natural materials into building components and products is _____.

    a. efficient energy

    b. conserved energy

    c. intense energy

    d. embodied energy

    Reference _____

7. Replacing 15 – 40 percent of portland cement in concrete with which material will result in increased concrete strength and provide environmental benefits?

    a. flyash

    b. sodium chloride

    c. fillers

    d. gypsum

    Reference _____

8. Double-pane, low-e windows with a thermal break spacer can reduce about_____ of the total house heat loss in the winter.

     a. 10 – 15 percent

     b. 15 – 20 percent

     c. 15 – 40 percent

     d. 20 – 40 percent

     Reference _____

9. Electrical energy consumption can be reduced by using _____.

     a. lighting controls

     b. low-voltage halogen lamps

     c. linear and compact fluorescent lamps

     d. any of the above

     Reference _____

10. Cotton insulation is considered a green building product, but it must be treated with borate for _____.

     a. increased R-value

     b. fire and insect protection

     c. mold resistance

     d. all of the above

     Reference _____

11. Which hot water circulation systems can result in 20 – 30 percent reduction in water use?

     a. Ground source

     b. Demand-controlled

     c. Preset or programmable

     d. Solar powered

     Reference _____

12. Recent cost analyses show that green building typically costs _____ more than conventional construction.

    a. 20 – 25 percent more

    b. 20 – 25 percent less

    c. 2 – 3 percent more

    d. 2 – 3 percent less

    Reference _____

13. To encourage environmentally responsible construction, some local communities offer incentives such as _____.

    a. lower permit fees

    b. expedited plan review

    c. public recognition

    d. any of the above

    Reference _____

14. Which of the following is a standard that establishes the principles and framework for life-cycle assessment?

    a. ISO 14040

    b. LEED (Leadership in Energy and Environmental Design)

    c. ASTM E2399-05

    d. Energy Star

    Reference _____

15. Code provisions pertaining to indoor air quality are likely to be found in the _____.

    a) *International Energy Conservation Code* (IECC)

    b) *International Mechanical Code* (IMC)

    c) *International Plumbing Code* (IPC)

    d) *International Mechanical Code* (IMC) and *International Building Code* (IBC)

    Reference _____

True/false

1. Greenhouse gases include carbon dioxide, nitrous oxide and methane.

    _____

    Reference _____

2. Site effectiveness, energy efficiency, water efficiency and indoor air quality are all considered green building practices.

    _____

    Reference _____

3. Life-cycle assessment does not include resource extraction and manufacturing processes.

    _____

    Reference _____

4. I-joists are lightweight and a more efficient use of forest resources than joists made of solid-sawn lumber.

    _____

    Reference _____

5. In establishing the energy intensity of building materials, the lower the number the less energy required to produce the material.

    _____

    Reference _____

6. The type of paint used and carpet installed can impact the indoor air quality of a building.

    _____

    Reference _____

7. An energy recovery ventilator can save 20 – 30 percent of energy normally required to heat or cool a building.

    _____

    Reference _____

Completion

1. In the United States, more than_____of electricity is used to construct and operate buildings.

    Reference _____

2. To understand sustainability, one must also understand _____.

    Reference _____

3. It's estimated that _____ of all material resources are used in the construction of buildings.

    Reference _____

4. _____insulation is made from recycled newspaper.

    Reference _____

5. Used for landscape irrigation and toilet/urinal flushing, _____ promote sustainability.

    Reference _____

6. Additional costs associated with green building are usually recovered through .

    Reference _____

7. A residential green building standard is being developed by the _____ and the ————————————.

    Reference _____

8. LEED is a national building rating system designed to _____.

    Reference _____

Answer Keys

Chapter 1

Multiple Choice

1.	c	Page 10,	Authority for Enforcing Codes
2.	c	Page 7,	Early Controls in the United States
3.	b	Page 13,	The Growth of Building Regulations
4.	c	Page 18,	ICC Evaluation Service
5.	b	Page 19,	International Accreditation Service
6.	a	Page 4,	The Chicago Fire
7.	b	Page 2,	introductory paragraphs
8.	d	Page 11,	Authority for Enforcing Codes
9.	b	Page 6,	Early Controls in the United States
10.	d	Page 8,	Where Are We Now?
11.	d	Page 20,	International Accreditation Service
12.	a	Page 15,	ICC Evaluation Service
13.	c	Page 25,	Summary
14.	d	Page 22,	Accreditation Process for Inspection Agencies
15.	c	Page7,	Early Controls in the United States

True/false

1.	T	Page 25,	Summary
2.	T	Page 11,	Scope of Codes
3.	F	Page 6,	Early Controls in the United States
4.	F	Page 2	introductory paragraphs
5.	F	Page 16,	ICC Evaluation Service

Completion

1. Materials, products and components

> Page 15, ICC Evaluation Service

2. Building Officials Conference of America, Pacific Coast Building Officials Conference, Southern Building Code Congress

> Page 6, Early Controls in the United States

3. Design, construction, use, occupancy, maintenance

> Page 10, Intent of Codes

4. Conflagration, earth tremors, carbon monoxide poisoning

> Page 14, The Growth of Building Regulation

5. Any two of the following four:
   economics
   lack of physical facilities
   absence of direct communication to decision making level
   inability of building official to provide convincing argument to support the need

> Page 12, Inadequately Staffed building Department

Chapter 2

Multiple Choice

1.	d	Page 28,	second paragraph
2.	a	Page 37,	Function of the FHA
3.	d	Page 29,	Performance Codes vs. Prescriptive Codes
4.	a	Page 38,	National Institute of Standards and Technology
5.	b	Page 33,	Governmental Consensus Process in Code Writing
6.	b	Page 33,	Code Adoption
7.	d	Page 33,	Governmental Consensus Process in Code Writing
8.	a	Page 32,	Code Changes and Reforms
9.	c	Page 37,	Function of the FHA
10.	c	Page 40,	Summary
11.	a	Page 39,	Organization and Functions
12.	c	Page 29,	Performance Codes vs. Prescriptive Codes
13.	d	Page 28,	second paragraph
14.	b	Page 35,	Governmental Consensus Process in Code Writing
15.	d	Page 38,	National Institute of Standards and Technology

True/false

1. F Page 38, National Institute of Standards and Technology
2. F Page 33, Governmental Consensus Process in Code Writing
3. T Page 34, Governmental Consensus Process in Code Writing
4. F Page 28, Second paragraph
5. T Page 33, Code Adoption

Completion

1. nonregulatory agency

   Page 38, National Institute of Standards and Technology

2. Any two of the following:
   references to other published codes
   generalization of a specific issue
   performance standards that lead to specifications
   use of an appendix
   use of the term accepted engineering practices

   Page 28, Performance Codes vs. Prescriptive Codes

3. complies with the intent of the provisions of the code
   at least equivalent of that described in this code in quality, strength, effectiveness, fire resistance, durability and safety

   Page 29, Performance Codes vs. Prescriptive Codes

4. Any two of the following:
   basic measurements and standards
   material measurements and standards
   technological measurements and standards
   transfer of technology

   Page 38 National Institute of Standards and Technology

5. Any three of the following:
   open
   transparent
   balanced
   achieves consensus
   complies with due process
   contains an appeal process
   meets federal law

   Page 35, Governmental Consensus Process in Code

Chapter 3

Multiple Choice

1. b   Page 52,   CABO and the Disappearing Differences in the Legacy Model Codes

2. c   Page 44,   The Development of Model Codes

3. a   Page 45,   The Development of Model Codes

4. d   Page 60,   The Courtship

5. c   Page 64,   Looking Forward

6. d   Page 42,   introductory paragraphs

7. a   Page 48,   ICBO and the Uniform Building Code

8. b   Page 48,   ICBO and the Uniform Building Code

9. d   Page 55,   The Vision

10. c   Page 54,   The Formation of the International Code Council

11. d   Page 54,   The Formation of the International Code Council

12. c   Page 60,   The Courtship

13. b   Page 65,   Summary

14. a   Page 45,   The Role of Building Officials

15. d   Page 44,   The Development of Model Codes

True/false

1. T   Page 45,   The Development of Model Codes

2. T   Page 59,   The Courtship

3. F   Page 64,   People Helping People

4. T   Page 42,   The Development of Model Codes

5. F   Page 45,   The Role of Building Officials

Completion

1. Any two of the following:

    Review three model codes to eliminate conflicts re: means of egress and type of construction.
    Develop specific code change proposals for each code to eliminate conflicts with respect to egress and type of construction.
    Develop specific proposed code changes to eliminate conflicts in other code areas.
    All code change proposals to eliminate conflict would be processed through the code change process for each participating organization.

    Page 53, Board for the Coordination of the Model Codes (BCMC)

2.   Any two of the following:

Business Growth
Core Functions
Customer Value
Policy and Thought Leadership
Social Responsibility/Public Visibility
Proper staffing needs

       Page 64/65, Looking Forward

3.   Pacific Coast Building Officials Conference
     International Conference of Building Officials

       Page 46, The Role of the Building Official

4.   Any two of the following:

Finance and Operations
Government Relations
Human Resources
Information Technology
Member Services
Technical Services
Business and Product Development

       Page 62, ICC Departments and Subsidiaries

5.   Any two of the following:

maintaining the intent of established provisions that adequately protected the public health, safety and welfare;
provisions that did not necessarily increase construction costs;
provisions that did not restrict the use of new materials, products or methods of construction; and
provisions that did not give preferential treatment to particular types or classes of materials, products or methods of construction

       Page 56, Merging Codes

Chapter 4

Multiple Choice

1.	d	Page 74,	ASTM International
2.	c	Page 68,	History and Development of Specific International Codes
3.	a	Page 81,	Underwriters Laboratories Inc.
4.	b	Page 83,	FM Global
5.	a	Page 68,	History and Development of Specific International Codes
6.	b	Page 77,	American National Standard Institute
7.	d	Page 86,	CSA America, Inc.
8.	b	Page 69,	International Building Code

9.	c	Page 75,	National Institute of Standards and Technology
10.	a	Page 69,	International Building Code
11.	d	Page 76,	American National Standards Institute
12.	b	Page 73,	The Standards Organizations and Testing for Safety
13.	a	Page 73,	ICC Performance Code for Buildings and Facilities
14.	b	Page 82,	The American Gas Association
15.	c	Page 77,	International Organization for Standardization

True/false

1.	F	Page 76,	American National Standards Institute
2.	T	Page 68,	History and Development of Specific International Codes
3.	T	Page 71,	International Plumbing Code
4.	F	Page 74,	ASTM International
5.	T	Page 80,	The National Fire Protection Association

Completion

1. Any two of the following:
   Plastic Plumbing Systems
   Drinking Water Additive
   Food Equipment

   Page 84/85, NSF International (formerly National Sanitation Foundation)

2. Detached one- and two-family dwellings and multiple single-family dwellings not more than three stories above-grade in height with separate means of egress.

   Page 72, *International Residential Code* (IRC)

3. Any two of the following: Safe, healthy, efficient, and accessible buildings and facilities.

   Page 68, History and Development of Specific International Codes

4. Balanced

   Page 68/69 History and Development of Specific International Codes

5. Any two of the following:

   To protect health, safety and welfare;
   to not unnecessarily increase construction costs;
   to not restrict the use of new materials, products or methods of construction;
   to not give preferential treatment to particular types or classes of materials, products, or methods of construction.

   Page 68, History and Development of Specific International Codes

Chapter 5

Multiple Choice

1.	c	Page 104,	Japan
2.	b	Page 109,	Mexico: The Current Project
3.	d	Page 120,	Spain
4.	b	Page 99,	Objective Based Codes
5.	a	Page 95,	Canada
6.	a	Page 115,	The Nordic Committee on Building Regulations
7.	d	Page 93,	Australian Building Code Board
8.	a	Page 101,	Role of the Canadian code Centre
9.	d	Page 93,	The Building Regulatory Review Task Force
10.	c	Page 117,	Norway
11.	d	Page 126,	International Accreditation Service
12.	b	Page 103,	England and Wales
13.	c	Page 113,	New Zealand
14.	a	Page 121,	Sweden
15.	d	Page 117,	Norway

True/false

1.	F	Page 104,	Japan
2.	T	Page 125,	International Accreditation Service
3.	F	Page 96,	Model National Code Documents
4.	F	Page 98,	Scope and Application of the Model Codes
5.	T	Page 102,	England and Wales

Completion

1. general provisions
   building codes
   zoning codes

   Page 104, Japan

2. enabling legislation
   a building code or regulation
   an enforcement mechanism

   Page 90, Overview of Building Regulatory Systems

3. Objectives

    Functional Statements
    Performance Requirements
    Deemed to Satisfy or Verification Methods

            Page 94, The Performance BCA

4. Follow prescriptive provisions.
    Follow provided verification methods
    Prove that alterative methods or materials satisfy the performance based provisions.

            Page 106, Japan

5. Tropical Cyclone North
    Earthquake Prone Areas
    Snow areas in New South Wales, Victoria, and Tasmania
    Bushfire-prone areas of the southeast and far southwest

            Page 92, the Building Code of Australia

Chapter 6

Multiple Choice

1. b   Page 140,   Additional Fees

2. b   Page 151,   Special Inspector

3. c   Page 158,   Small Jurisdictions: Special Needs and Issues

4. d   Page 151,   Special Inspector

5. a   Page 146,   Plan Checking

6. d   Page 149,   Inspection

7. d   Page 130,   Organization of the Department

8. c   Page 130,   Organization of the Department

9. d   Page 132,   External Assistance

10. a   Page 135,   Managing the Department

11. d   Page 138,   Fixed Fee vs. Actual Costs

12. a   Page 152,   Insurance Services Office

13. c   Page 152,   Insurance Services Offices

14. d   Page 154,   On-Site Evaluation

15. c   Page 135,   Managing the Department

True/false

1. F   Page 147,   Plan Checking

2. T   Page 141,   Enterprise Funds

3.	F	Page 147,	Plan Checking
4.	T	Page 148,	Inspection
5.	T	Page 132,	How Much Discretion?

Completion

1. Any two of the following:

   prestressed concrete
   reinforced concrete
   spray-applied fireproofing
   masonry
   steel

   > Page 151, Special Inspectors

2. Permit issuance
   Plan examination
   Inspections

   > Page 130, The Duty of the Building Department

3. Personal appearance
   Manners
   Communications skills

   > Page 135, Managing the Department

4. Assesses the good and bad equally.

   > Page 138, Fixed Fee vs. Actual Costs

5. Any three of the following:

   Topography
   Population mix
   Types of construction
   Service goals
   Department budget and staffing levels
   Quality systems
   Appeals

   > Page 153, International Accreditation Service

Chapter 7

Multiple Choice

1.	c	Page 181,	Planning and Zoning Commission
2.	a	Page 170,	Subdivision Engineering Process
3.	c	Page 173,	Emergence of Regulatory Departments
4.	c	Page 182,	Building Appeals Board

5.	d	Page 176,	Planning
6.	a	Page 176,	Public Works
7.	b	Page 178,	Chamber of Commerce/Economic Development
8.	a	Page 177,	Engineering
9.	d	Page 169,	Zoning Process
10.	a	Page 177,	Building Inspections
11.	c	Page 178,	Building Inspection
12.	a	Page 180,	Design Consultants
13.	d	Page 182,	Zoning board of Adjustment
14.	a	Page 169,	Zoning Process
15.	c	Page 171,	Building Permit Process

True/false

1.	T	Page 195,	Summary
2.	F	Page 179,	Fire Prevention
3.	T	Page 179,	Design Consultants
4.	T	Page 176,	Planning
5.	F	Page 181,	The General Contractor and Developer

Completion

1. to serve the public by ensuring that the life safety and community welfare contemplated by its various codes and ordinances becomes an accomplished fact.

    Page 174, Development Departments

2. zoning, subdivision engineering, building construction

    Page 168, The Practice

3. market conditions and economic trends,
   environmental regulations and political governance

    Page 183, Trends

4. responsibility for protecting the public,
   advancing the economic development and urban growth for communities

    Page 166, introductory paragraphs

5. Any two of the following:

    project management,
    communications with customers,
    expediting the review process
    process improvement

    Page 186, Growth and Demand

Chapter 8

Multiple Choice

1.	b	Page 209,	Leadership
2.	a	Page 212,	Encourage Teamwork and Cooperation
3.	c	Page 219,	Delegation
4.	d	Page 205,	Duties
5.	a	Page 211,	Keep an Open Mind
6.	c	Page 222,	Effective Speaking
7.	d	Page 239,	Summary
8.	a	Page 205,	Modifications
9.	b	Page 209,	Leadership
10.	a	Page 225,	Written Policy Statements
11.	c	Page 198,	first paragraph
12.	d	Page 199,	Industrial Economy
13.	b	Page 228,	The Code of Ethics
14.	a	Page 239,	Summary
15.	c	Page 204,	Duties

True/false

1.	T	Page 215,	Discipline and Morale
2.	F	Page 205,	Duties
3.	F	Page 211,	Accept Responsibility
4.	F	Page 213,	Leadership Traits
5.	T	Page 220,	Develop Employee's Potential

Completion

1. Any two of the following:

    planning
    organizing
    leading
    controlling

Page 209, Leadership

2.   Any two of the following:

Identify the problem
Discuss the reason for the question
Listen to every opinion
Clarify the intent or procedure in a policy statement

Page 226, Written Policy Statements

3.   Any two of the following:

convey an interest in staff opinions
show value for their contribution
give staff opportunity to be heard
get a view of impending trouble

Page 227, Keeping the Staff Informed

4.   Ask for an advanced list of questions
Ask for the right to review the interview before publishing
Try to get to know the reporter before the interview

Page 236, Interaction With the Media

5. Any two of the following:

Keep an open mind
Celebrate the Building Department's achievements
Accept responsibility
Encourage teamwork and cooperation
Possess and use leadership traits

Page 210–215, Rules of Conduct

Chapter 9

Multiple Choice

1.	c	Page 245,	Community Leadership
2.	a	Page 251,	Promoting the Community Future
3.	c	Page 243,	Chapter introductory paragraphs, last paragraph
4.	b	Page 247,	Excellence in Management
5.	b	Page 244,	Government Manager
6.	c	Page 243,	Chapter introductory paragraphs, last paragraph
7.	d	Page 243,	Government Manager
8.	a	Page 246,	Sources of Conflict
9.	b	Page 249,	Excellence in Management
10.	d	Page 251,	Managing the Manager's Life

11.	a	Page 248,	Excellence in Management
12.	b	Page 250,	Managing Public Policy
13.	d	Page 249,	second paragraph
14.	d	Page 248,	Excellence in Management
15.	c	Page 246,	Sources of Conflict

True/false

1.	F	Page 251,	Promoting the community future
2.	F	Page 247	Excellence in Management
3.	F	Page 250,	Promoting the Community Future
4.	F	Page 252,	Summary
5.	T	Page 249,	Excellence in Management

Completion

1. Any two of the following:

   government functions
   resources
   responsibilities

   Page 242, Chapter introductory paragraphs, 2nd paragraph

2. Any two of the following:

   the changing environment
   the needs of the public interest
   constraints faced by the community

   Page 244, "Government Manager"

3. Ineffective application of code
   Misapplication of standards

   Page 246. "Sources of Conflict

4. Any three of the 18 listed on page 242, such as:

   Being responsible sometimes means pissing people off
   Never neglect details
   When everyone's mind is dulled or distracted, the leader must be doubly vigilant
   The commander in the field is always right and rear echelon is wrong, unless proved otherwise

   Page 242, introductory paragraphs

5. Productivity and service delivery

   Page 244, third paragraph, Government Manager

Chapter 10

Multiple Choice

1.	b	Page 270,	Department Procedure Manual
2.	a	Page 264,	The Supervisor
3.	d	Page 254,	Chapter introductory paragraphs
4.	a	Page 255,	Integrity
5.	b	Page 258,	Leadership
6.	c	Page 258,	Creating the Will to Work
7.	a	Page 263,	Performance Review
8.	c	Page 272,	Training
9.	c	Page 274,	New Employees
10.	b	Page 278,	Certification Programs
11.	d	Page 275,	Other Training Considerations
12.	a	Page 273,	Responsibility for Training
13.	a	Page 279,	Summary
14.	c	Page 259,	Creating the Will to Work
15.	d	Page263,	The Performance Review

True/false

1.	F	Page 257,	Leadership
2.	T	Page 256,	Attitudes
3.	T	Page 273,	Training
4.	T	Page 259,	Creating the Will to Work
5.	F	Page 272,	Department Procedures Manual: Notification of the Trades

Completion

1. self-motivation, temporary incentive, fear

    Page 277, Certification Programs

2. a complete explanation

    Page 255, Integrity

3. service oriented, helpful, friendly

    Page 257, Attitudes

4. employees as human beings

    Page 258, Creating the Will to Work

5. emotional and psychological income

> Page 259, Creating the Will to Work

Chapter 11

Multiple Choice

1.	b	Page 282,	Department Structure
2.	d	Page 283,	Organization Chart
3.	d	Page 283,	Identification of Key Functions
4.	a	Page 284,	Identification of Key Functions
5.	c	Page 285,	Performance Measures and Indicators
6.	a	Page 285,	Quantitative Standards
7.	c	Page 286,	Determining Staffing
8.	d	Page 287,	Inspectors
9.	b	Pages 287–290,	Inspectors
10.	a	Pages 291–292,	Workload Determined by Revenue
11.	c	Page 294,	Selecting Personnel
12.	b	Page 294,	Selecting Personnel
13.	d	Pages 295–297,	Communicating
14.	a	Page 297,	Assignment of Functions
15.	d	Page 303,	Summary

True/false

1.	F	Page 282,	Department Structure
2.	F	Page 303,	Summary
3.	F	Page 290,	Certified Plan Examiners
4.	T	Page 283,	Key Functions and Page 297, Assignment of Functions
5.	F	Page 302,	Assignment of Functions

Completion

1. quantitative performance measure

> Page 285, Quantitative Standards

2. time and motion study

> Page 287, Inspectors

3. opinion

   Page 294, Selecting Personnel

4. Answering phone calls
 Answering code questions at public counter
 Attending community meetings

   Page 293, Identification of Revenue/Non-Revenue Functions

5. fair, factual, firm, final

   Page 296, Communicating

Chapter 12

Multiple Choice

1.	c	Page 306,	Doing the Right Thing
2.	d	Page 306,	Doing the Right Thing
3.	a	Page 307,	First paragraph
4.	b	Page 307,	Third paragraph
5.	a	Page 308,	Second paragraph
6.	c	Page 309,	Third paragraph
7.	d	Page 309,	Third paragraph
8.	a	Page 311,	Second paragraph
9.	d	Pages 311–312,	Permit Processing and Issuance
10.	c	Page 318,	#5 – Counter Checking
11.	b	Page 320,	Second paragraph
12.	a	Page 322,	Valuation Disputes
13.	c	Page 323,	Belief in the Importance of Codes
14.	c	Page 324,	Communication
15.	d	Page 326,	Further Suggestions

True/false

1.	F	Page 309, last paragraph
2.	T	Page 306, First paragraph
3.	T	Page 323, Belief in the Importance of Codes
4.	T	Page 324, Communication
5.	F	Page 325, The Resolution Center

Completion

1. the permit-issuance function

Page 326, Summary

2. application submittal
plan review or examination
permit fee collection and permit issuance

Pages 314–318, Example of a Step-by-Step Procedure

3. Being polite and informative

Page 308, first paragraph

4. An increase in their property taxes
Excessive permit fees

Page 319, Determining Valuation

5. Permit applications
Checklists and procedures for plan review
Checklists and procedures for inspection
Sample drawings
Local code amendments
Fee schedules
Informational handouts

Page 325, The Resolution Center

Chapter 13

Multiple Choice

1.	a	Page 328,	How is Technology Used?
2.	c	Page 328,	Why Implement a New Solution?
3.	c	Page 329,	What Constitutes "E-Permitting"?
4.	b	Page 330,	Types of Electronic Permitting Systems
5.	d	Page 330,	Available Functionality
6.	b	Page 331,	Second paragraph
7.	a	Page 332,	Third paragraph
8.	c	Page 334,	System Support and Management, #2-Service Contracts
9.	d	Page 336,	Create an Electronic Permitting Task Force
10.	d	Page 337,	Select the Type of System
11.	a	Page 338,	Select the Type of System
12.	d	Pages 339–340,	Establish an Implementation Team
13.	c	Page 340,	Implementation, #4-Database Migration
14.	a	Page 345,	Handwriting Scanning-Pen Technology
15.	d	Pages 347–348,	Can "Virtual" Inspections Beat the Clock?

True/false

1. T   Page 329,   Why Implement a New Solution?

2. T   Page 330,   Available Functionality

3. F   Page 332,   Third paragraph

4. F   Page 335,   Determine the Need

5. F   Page 340,   Implementation, #4-Database Migration

Completion

1. the department
   other agencies within the jurisdiction
   constituents

         Page 349, Summary

2. Any two of the following:

   reduce permitting time
   improve customer service and staffing efficiency
   enhance quality
   make operating funds more productive

         Page 328, How is Technology Used?

3. schedule inspections and obtain results
   track plan review

         Page 331, Available Functionality

4. Any three of the following:

   Statement of purpose
   Glossary of terms
   Description of expected results
   Technical and functional requirements
   Detailed process requirements

         Page 338, Prepare a Request for Proposals

5. the amount of inspector travel time

         Page 348, Third paragraph

Chapter 14

Multiple Choice

1. c   Page 352,   The Need for Records

2. c   Page 352,   The Need for Records

3. a   Pages 353–354,   Purification of Files

4.	d	Page 354,	Accuracy
5.	d	Page 355,	Inconsistent Practices
6.	b	Pages 356-358,	Vital Information; Valuation
7.	b	Page 369,	First paragraph
8.	a	Page 370,	Forms Revision
9.	d	Page 372,	Plans Check Report
10.	c	Page 376,	Form Letters
11.	c	Page 383,	Certificate of Occupancy
12.	a	Page 387,	Value of Report
13.	d	Page 387–388,	Reporting Format
14.	b	Pages 390–392,	Diagrams
15.	c	Page 396,	Summary

True/false

1.	F	Page 352,	The Need for Records
2.	F	Page 353,	Nonpublic Records
3.	T	Page 359,	Signatures
4.	T	Page 362,	Single Building Permit
5.	F	Page 385,	Certificate Designs

Completion

1. permanent, transitory
   public, nonpublic

   Pages 352–353, The Need for Records

2. Reduced plan review time
   Clear, precise wording

   Page 370, Plans Review Reports, General

3. Request for inspection
   No admission tag
   Inspection record
   Dangerous building notice
   Daily inspection report
   Correction notices

   Pages 376–383, Inspection Records

4. Lends dignity to a business
   More likely to be displayed
   Helps publicize role of building inspector

Page 385, Certificate Designs

5. Number of permits issued
    Number of buildings completed per type and use
    Number of inspections made
    Building valuation
    Revenue by permit type
    Revenues versus expenses

    Page 387, Reporting Format

Chapter 15

Multiple Choice

1.	a	Page 398,	First and second paragraphs
2.	d	Page 399,	The Variety of Good Relations
3.	d	Page 400,	Controversies
4.	d	Page 401,	The General Public
5.	d	Page 402,	Publicity is Necessary
6.	a	Page 403,	How Much Service?
7.	c	Page 404,	Architects and Engineers
8.	c	Pages 403–405,	Architects and Engineers
9.	a	Page 407,	Prohibitions
10.	d	Page 409,	Approvals
11.	d	Pages 411–414,	Political Relations
12.	c	Pages 415–416,	Relationship between Fire Prevention & Building Regulations
13.	c	Page 417,	Media Relations
14.	b	Pages 420–423,	Complaint Handling
15.	d	Page 401,	Owner-Contractor Disputes

True/false

1.	F	Page 398,	The Variety of Good Relations
2.	T	Page 404,	Architects and Engineers
3.	T	Page 405,	Second paragraph
4.	F	Page 420,	Complaint Handling

5. F   Page 423,   Relationships with Other Building Officials

Completion

1. Fill out a complaint form
    Investigate and record findings
    Notify the complainant
    Close the case upon resolution

    Page 421, Recording & Monitoring Procedure

2. Patience

    Page 402, The Virtues of Patience

3. Press releases to the news media
    Speaking to business and community groups
    Promotional campaigns (i.e., Building Safety Week)

    Page 402, Publicity is Necessary

4. save time and wear and tear on deptartment personnel

    Page 421, Courtesy to Complainant

5. "tough but fair"

    Page 424, Summary

Chapter 16

Multiple Choice

1. c   Page 428,   introductory paragraphs, paragraph 1

2. d   Page 428,   introductory paragraphs, paragraph 2

3. b   Page 429,   Enforcement and the Building Official

4. d   Page 431,   Permitting Code Violations

5. a   Page 434,   Sovereign Immunity

6. c   Page 437,   Law and the Building Official

7. a   Page 441,   Inverse Condemnation

8. c   Page 443,   Covenants and Restrictions

9. c   Page 444,   Prohibitive Practice

10. d   Page 447,   Establishing a Legal Case

11. b   Page 453,   Consent

12. c   Pages 451–454   Right of Entry, Probable Cause, Consent, Public Areas

13. a   Page 457,   Negligence

14. b   Page 466,   Statute of Limitations

15. c   Page 451,   Right of Entry

True/false

1. T   Page 431,   Permitting Code Violations
2. F   Page 428,   Duties of the Building Official
3. T   Page 433,   Doctrine of Preemption
4. T   Page 435,   Special Districts
5. F   Page 438,   Legal Terminology

Completion

1. easement

        Page 443, Easement

2. police power

        Page 445, Enforcement

3. state action

        Page 451, Right of Entry

4. voluntarily

        Page 453, Consent

5. Fixed period of time

        Page 466, Statute of Limitations

Chapter 17

Multiple Choice

1. c   Page 471,   First paragraph
2. b   Page 471,   Disaster Mitigation Strategies and Preparedness
3. a   Page 474,   Land-Use Planning
4. d   Page 475,   Hazard Control
5. d   Page 475,   Mitigation Programs
6. a   Page, 481,   Tornado/Hurricane Safe Rooms
7. d   Pages 486–487, Structural Retrofits
8. a   Page 493,   Antiterrorism and Counterterrorism
9. c   Page 494,   Third paragraph
10. b   Page 471,   A Brief History of Hazard Mitigation
11. b   Page 481,   Residential Initiatives
12. d   Page 493,   Antiterrorism and Counterterrorism

13.	a	Page 492,	Ice and Snow
14.	d	Page 487,	National Earthquake Hazards Reduction Program (NEHRP)
15.	a	Page 485,	Nonstructural retrofits

True/false

1.	T	Page 470,	First paragraph
2.	T	Page 471,	Second paragraph
3.	T	Page 473,	Design and Construction
4.	T	Page 476,	Hazard Mitigation Grant Program
5.	F	Page 484,	Earthquake Mitigation Measures

Completion

1. Design and construction
   Land-use planning
   Organizational planning
   Hazard control

   Page 472, Mitigation Tools

2. undeveloped areas
   areas where there hasn't been an investment in public infrastructure

   Page 474, Land-Use Planning

3. Any three of the following:
   Public Assistance Program
   Hazard Mitigation Grant Program
   Flood Mitigation Assistance Program
   Pre-disaster Mitigation Grant Program
   Disaster Resistant University Program

   Page 475, Mitigation Programs

4. Response and Recovery
   Planning and Preparedness
   Mitigation

   Page 479, National Hurricane Program

5. landscaping wisely
   using nonflammable construction materials

   Page 491, entire page

Chapter 18

Multiple Choice

| 1. | d | Page 498, | Introduction |
| 2. | b | Page 499, | The Housing Code |

3.	c	Page 500,	Background
4.	d	Page 501,	Background
5.	b	Page 502,	third paragraph
6.	a	Page 504,	Housing and Property Maintenance Enforcement Programs
7.	d	Page 505,	Proactive Approach
8.	a	Page 507,	Assessment or Benchmarks
9.	d	Page 509,	Performance Measures
10.	b	Page 512,	Housing Inspections
11.	d	Pages 513–514,	Abatement of Substandard Buildings
12.	c	Pages 513–514,	Abatement of Substandard Buildings
13.	a	Page 519,	Due Process Requirements
14.	c	Page 521,	Forms and Form Letters
15.	c	Page 523,	Tips for Effective Code Enforcement

True/false

1.	T	Page 499,	The Housing Code
2.	F	Pages 499–500,	The Housing Code
3.	T	Page 503,	second paragraph
4.	T	Page 505,	Reactive Programs
5.	F	Page 517,	Single-Family Rental Inspection Program

Completion

1. the Housing Act of 1949

    Pages 498–499, Brief History of Housing

2. to establish minimum standards essential to make dwellings safe, sanitary and fit for human habitation

    Pages 498–499, Brief History of Housing

3. how one builds a structure
    how one healthfully lives in and maintains a structure

    Pages 500–501, Background

4. Assessment or benchmarks
    Strategic plan
    Program initiatives
    Performance measures

    Pages 505–506, Proactive Approach

5. use state or municipal codes, if available, to obtain a warrant for access

    Page 523, #10, Tips for Effective Code Enforcement

Chapter 19

Multiple Choice

1.	d	Page 526,	first paragraph
2.	c	Page 527,	first paragraph
3.	c	Page 527,	Table 19-1
4.	b	Page 532,	Table 19-3
5.	a	Page 533,	Economic Constraints
6.	a	Page 537,	Building Code Barriers to Rehab
7.	b	Page 540,	Building Codes Best Practices for Rehab
8.	d	Page 541,	Building Codes Best Practices for Rehab
9.	c	Page 541,	Building Codes Best Practices for Rehab
10.	c	Page 547,	Historic Preservation: Background
11.	d	Page 547,	Historic Preservation: Background
12.	a	Pages 547–548,	Historic Preservation: Background
13.	b	Page 548,	Historic Preservation Contributes to Housing Rehabilitation
14.	a	Page 548,	Historic Preservation Contributes to Housing Rehabilitation
15.	d	Pages 552–553,	Table 19-9

True/false

1.	T	Page 526,	first paragraph
2.	T	Page 528,	second paragraph
3.	F	Page 529,	Analytic Framework
4.	F	Page 541,	Second paragraph
5.	T	Page 547,	Historic Preservation: Background

Completion

1. 70–71 percent

   Page 533, Economic Constraints

2. often reinforcing

   Page 535, The Development, Construction and Occupancy Challenges

3. technically problematic
   expensive

   Page 537, Building Code Barriers to Rehab

4. Section 106 of the National Historic Preservation Act of 1966

   Page 547, Historic Preservation: Background

5. protective benefits

Page 548, Historic Preservation Contributes to Housing Rehabilitation

Chapter 20

Multiple Choice

1.	b	Page 556,	Introduction
2.	d	Page 557,	Background
3.	a	Page 558,	Differences: Rehab and New Building Construction
4.	b	Page 559,	Administrative and Technical Process for Rehab
5.	c	Page 557,	Background
6.	a	Page 557,	Background

True/false

1.	F	Page 556,	Definitions
2.	F	Page 556,	Definitions
3.	F	Page 557,	Background
4.	T	Page 558,	second paragraph
5.	T	Page 558,	Differences: Rehab and New Building Construction
6.	T	Page 559,	Administrative and Technical Process for Rehab

Chapter 21

Multiple Choice

1.	a	Page 562,	first paragraph
2.	b	Page 562,	first paragraph
3.	b	Page 562,	The Environmental Impact of Buildings
4.	c	Page 563,	Sustainability and Green Building
5.	d	Page 563,	Sustainability and Green Building
6.	d	Page 565,	Building Materials
7.	a	Page 567,	Table 21-3, Flyash
8.	c	Page 569,	Table 21-3, Windows
9.	d	Page 570,	Table 21-3, Lighting
10.	b	Page 568,	Table 21-3, Cotton insulation
11.	b	Page 569,	Table 21-3, Hot water circulation pump

12.	c	Page 570,	first paragraph
13.	d	Page 572,	first paragraph
14.	a	Page 574,	first paragraph
15.	d	Page 574,	Figure 21-3

True/false

1.	T	Page 562,	The Environmental Impact of Buildings
2.	T	Page 563,	Sustainability and Green Building
3.	F	Page 565,	Building Materials
4.	T	Page 567,	Table 21-3, I-joist
5.	T	Page 565,	Building Materials
6.	T	Page 566,	Interior Finishes
7.	T	Page 570,	Table 21-3, Economizer

Completion

1. 60 percent

    Page 562, The Environmental Impact of Buildings

2. the interdependence of the human and natural environments

    Page 563, Sustainability and Green Building

3. more than 40 percent

    Page 565, Building Materials

4. Cellulose

    Page 568, Table 21-3, Cellulose insulation

5. graywater and rainwater reuse systems

    Page 569, Table 21-3, Plumbing

6. savings in utility, operating and maintenance costs

    Page 570, First paragraph

7. International Code Council (ICC)
   National Association of Homebuilders (NAHB)

    Page 575, Summary

8. accelerate the development and implementation of sustainable building practices

    Page 572, Second paragraph

Increase Your Legal Knowledge

A

B

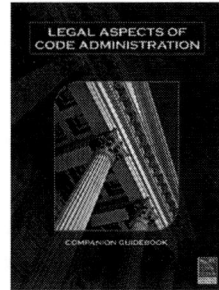

C

A–B: LEGAL ASPECTS OF CODE ADMINISTRATION
This in-depth explanation of the profession of administering and enforcing building codes is organized in a logical sequence with explanation of legal terminology. It serves as a refresher for those preparing to take the legal module of the ICC Certified Code Official examinations.

A: AUDIOBOOK (Unabridged, 6 CD set, approximately 6.5 hours)
#7807S

B: TEXTBOOK (160 pages)
#1007S01

BUY BOTH AND SAVE!
#7807BN

C: LEGAL ASPECTS OF CODE ADMINISTRATION COMPANION GUIDEBOOK
Serves as a companion to the *Legal Aspects of Code Administration* textbook. Assignment boxes throughout the guidebook coordinate use of both books and gives instruction for:
- Matching terms and definitions
- Reading textbook pages
- Reading summaries
- Completing self-evaluations
- Evaluating case studies
References to the textbook appear in bold-faced type. (200 pages)
#1008S01

FIRST PREVENTERS' SAFETY CHECKLISTS
Educate the public about five important safety topics: Decks, Porches and Balconies; Smoke Alarms; Pools, Spas and Hot Tubs; Outdoor Cooking Grills; and Building Permits. The package contains 25 copies of each pocket-sized checklist. (125 checklists)
#7327S

ORDER YOURS TODAY! 1-800-786-4452 | www.iccsafe.org/store

ICC INTERNATIONAL CODE COUNCIL®

People Helping People Build a Safer World™

09-01589

ICC EVALUATION SERVICE

Innovative Building Products:

The Code Requirement

The International Building Code® Section 104.11 allows for the use of alternate building products. However, it requires building officials to verify that the proposed design is satisfactory and complies with the intent of the code. Specifically, the material, method or work offered must be at least the equivalent of that prescribed in the code in quality, strength, effectiveness, fire resistance, durability and safety.

What's in an ICC-ES Evaluation Report

Evaluation reports from ICC Evaluation Service® are the most preferred resource used by code officials to verify that new and innovative building products comply with code requirements. The evaluation reports provide information about what code requirements or acceptance criteria were used to evaluate the product, how the product should be installed to meet the requirements, how to identify the product, and much more. ES Reports are divided into eleven major areas.

1 **CSI Division Number**—ICC-ES Evaluation Reports, and the building products represented in them, are organized according to the Construction Specifications Institute's (CSI) Masterformat system.

2 **Report Holder**—The name and address of the company or organization that has applied for the Evaluation Report.

3 **Evaluation Subject**—The specific product(s) covered by the report.

4 **Evaluation Scope**—The code(s) that were used to evaluate the product.

5 **Properties Evaluated**—A brief description of the properties the product was evaluated against such as fire resistance and wind resistance. This section also shows if the product can be used for structural purposes.

6 **Uses**—Identifies the scope of the Evaluation Report and relates the product evaluated to code provisions.

7 **Description**—Provides a general description of the product and its features, such as length, thickness, etc.

8 **Installation**—Identifies general and often specific requirements to help the inspector ensure the product is installed properly according to the code requirements or acceptance criteria.

9 **Conditions of Use**—Statement that the product, as described in the Evaluation Report, complies with or is a suitable alternative to the requirements of the applicable code and a list of conditions under which the report is issued.

10 **Evidence Submitted**—Data (i.e. test reports, calculations, installation instructions) that was used in evaluating the product.

11 **Identification**—Information that can be used to identify the product, including the manufacturer's name, product code, Evaluation Report number, etc.

Make sure they are up to code with ICC-ES Evaluation Reports

The ICC-ES Solution

ICC Evaluation Service® (ICC-ES®), a subsidiary of ICC®, was created to assist code officials and industry professionals in verifying that new and innovative building products meet code requirements. This is done through a comprehensive evaluation process that results in the publication of ICC-ES Evaluation Reports for those products that comply with requirements in the code or acceptance critiera. Today, more code officials prefer using ICC-ES Evaluation Reports over any other resource to verify products comply with codes.

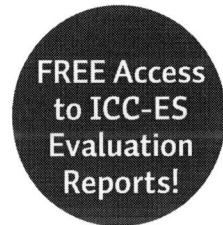

ICC EVALUATION SERVICE

Most Widely Accepted and Trusted

ICC-ES Evaluation Report

ESR-4802

Issued March 1, 2008

This report is subject to re-examination in one year.

www.icc-es.org | 1-800-423-6587 | (562) 699-0543   *A Subsidiary of the International Code Council®*

1 DIVISION: 07—THERMAL AND MOISTURE PROTECTION
Section: 07410—Metal Roof and Wall Panels

2 REPORT HOLDER:

ACME CUSTOM-BILT PANELS
52380 FLOWER STREET
CHICO, MONTANA 43820
(808) 664-1512
www.custombiltpanels.com

3 EVALUATION SUBJECT:

CUSTOM-BILT STANDING SEAM METAL ROOF PANELS: CB-150

4 1.0 EVALUATION SCOPE
Compliance with the following codes:
- 2006 *International Building Code®* (IBC)
- 2006 *International Residential Code®* (IRC)

5 Properties evaluated:
- Weather resistance
- Fire classification
- Wind uplift resistance

6 2.0 USES
Custom-Bilt Standing Seam Metal Roof Panels are steel panels complying with IBC Section 1507.4 and IRC Section R905.10. The panels are recognized for use as Class A roof coverings when installed in accordance with this report.

7 3.0 DESCRIPTION
3.1 Roofing Panels:
Custom-Bilt standing seam roof panels are fabricated in steel and are available in the CB-150 and SL-1750 profiles. The panels are roll-formed at the jobsite to provide the standing seams between panels. See Figures 1 and 3 for panel profiles. The standing seam roof panels are roll-formed from minimum No. 24 gage [0.024 inch thick (0.61 mm)] cold-formed sheet steel. The steel conforms to ASTM A 792, with an aluminum-zinc alloy coating designation of AZ50.
3.2 Decking:
Solid or closely fitted decking must be minimum $^{15}/_{32}$-inch-thick (11.9 mm) wood structural panel or lumber sheathing, complying with IBC Section 2304.7.2 or IRC Section R803, as applicable.

8 4.0 INSTALLATION
4.1 General:
Installation of the Custom-Bilt Standing Seam Roof Panels must be in accordance with this report, Section 1507.4 of the IBC or Section R905.10 of the IRC, and the manufacturer's

published installation instructions. The manufacturer's installation instructions must be available at the jobsite at all times during installation. The roof panels must be installed on solid or closely fitted decking, as specified in Section 3.2. Accessories such as gutters, drip angles, fascias, ridge caps, window or gable trim, valley and hip flashings, etc., are fabricated to suit each job condition. Details must be submitted to the code official for each installation.

4.2 Roof Panel Installation:
4.2.1 CB-150: The CB-150 roof panels are installed on roof shaving a minimum slope of 2:12 (17 percent). The roof panels are installed over the optional underlayment and secured to the sheathing with the panel clip. The clips are located at each panel rib side lap spaced 6 inches (152 mm) from all ends and at a maximum of 4 feet (1.22 m) on center along the length of the rib, and fastened with a minimum of two No. 10 by 1-inch pan head corrosion-resistant screws. The panel ribs are mechanically seamed twice, each pass at 90 degrees, resulting in a double-locking fold.

4.3 Fire Classification:
The steel panels are considered Class A roof coverings in accordance with the exception to IBC Section 1505.2 and IRC Section R902.1.

4.4 Wind Uplift Resistance:
The systems described in Section 3.0 and installed in accordance with Sections 4.1 and 4.2 have an allowable wind uplift resistance of 45 pounds per square foot (2.15 kPa).

9 5.0 CONDITIONS OF USE
The standing seam metal roof panels described in this report comply with, or are suitable alternatives to what is specified in, those codes listed in Section 1.0 of this report, subject to the following conditions:

5.1 Installation must comply with this report, the applicable code, and the manufacturer's published installation instructions. If there is a conflict between this report and the manufacturer's published installation instructions, this report governs.

5.2 The required design wind loads must be determined for each project. Wind uplift pressure on any roof area must not exceed 45 pounds per square foot (2.15 kPa).

10 6.0 EVIDENCE SUBMITTED
Data in accordance with the ICC-ES Acceptance Criteria for Metal Roof Coverings (AC166), dated October 2007.

11 7.0 IDENTIFICTION
Each standing seam metal roof panel is identified with a label bearing the product name, the material type and gage, the Acme Custom-Bilt Panels name and address, and the evaluation report number (ESR-4802).

Page 1 of 1

VIEW ONLINE NOW!
www.icc-es.org